Vietnam's Post-1975
AGRARIAN REFORMS

How local politics derailed socialist agriculture
in southern Vietnam

Vietnam's Post-1975 AGRARIAN REFORMS

How local politics derailed socialist agriculture in southern Vietnam

TRUNG DANG

Australian
National
University

PRESS

VIETNAM SERIES

ANU PRESS

Published by ANU Press
The Australian National University
Acton ACT 2601, Australia
Email: anupress@anu.edu.au
This title is also available online at press.anu.edu.au

A catalogue record for this book is available from the National Library of Australia

NATIONAL LIBRARY OF AUSTRALIA

ISBN(s): 9781760461959 (print)
          9781760461966 (eBook)

Cover design and layout by ANU Press. Cover image: Workers thresh rice in a village in Quảng Nam province, 4 July 2006, by Dr Ashley Carruthers, The Australian National University.

Contents

Figures

Maps

Tables

Abbreviations

BCHDBHCM	Ban Chấp Hành Đảng Bộ Huyện Chợ Mới (Chợ Mới Party Executive Committee)
BCHTU	Ban Chấp Hành Trung Ương (Central Executive Committee)
BCTNN	Ban Cải Tạo Nông Nghiệp (Committee for Agricultural Transformation)
BCTNNAG	Ban Cải Tạo Nông Nghiệp An Giang (An Giang Committee for Agricultural Transformation)
BCTNNMH	Ban Cải Tạo Nông Nghiệp Minh Hải (Minh Hải Committee for Agricultural Transformation)
BCTNNMN	Ban Cải Tạo Nông Nghiệp Miền Nam (Committee for Southern Agricultural Transformation)
CCTKCM	Chi Cục Thống Kê huyện Chợ Mới (Chợ Mới District Statistical Office)
CTKAG	Cục Thống Kê An Giang (An Giang Department of Statistics)
CTKQN	Cục Thống Kê Quảng Nam (Quảng Nam Department of Statistics)
ĐBCM	Đảng Bộ Chợ Mới (Chợ Mới party cell)
ĐCSVN	Đảng Cộng Sản Việt Nam (Vietnamese Communist Party)
HTX	*hợp tác xã* (collective)
LTTT	Land to the Tiller program
NLF	National Liberation Front
PST	production solidarity team (*tổ đoàn kết sản xuất*)

PTWCNC	production team working according to norms and contracts (*tổ sản xuất có định mức, khoán việc*)
QN-ĐN	Quảng Nam-Đà Nẵng province
SNNPTNTQN	Sở Nông Nghiệp Phát Triển Nông Thôn Quảng Nam (Quảng Nam Department of Agriculture and Rural Development)
TP	Thành Phố (City)
TU	*tỉnh ủy* (provincial party committee)
TUQN	Tỉnh Ủy Quảng Nam (Quảng Nam Party Committee)
UBNDTAG	Ủy Ban Nhân Dân Tỉnh An Giang (An Giang People's Committee)
VCP	Vietnamese Communist Party (*Đảng Cộng Sản Việt Nam*)
VND	Vietnamese dong (currency)

Key Vietnamese terms

Các hình thức tập dượt của hợp tác xã	Interim forms of collectives
Cách mạng ruộng đất	Land revolution
Cách mạng xã hôi chủ nghĩa	Socialist revolution
Cải tạo nông nghiệp	Agricultural transformation
Công điểm	Work-points
Công điền công thổ	Communal land
Điều chỉnh ruộng đất	Land redistribution (readjustment)
Hợp tác hóa	Collectivisation
Hợp tác xã	Collective
Hợp tác xã bậc cao	High-level collective
Hợp tác xã bậc thấp	Low-level collective
Khoán sản phẩm	Product contract system
Lúa mùa	Traditional rice
Lúa nổi	Floating rice
Lúa thần nông	High-yielding rice
Nông trường quốc doanh	State farm
Sản xuất lớn xã hội chủ nghĩa	Socialist large-scale production
Tập đoàn máy	Machinery unit
Tập đoàn sản xuất	Production unit
Tổ đoàn kết sản xuất	Production solidarity team
Tổ đổi công, vần công	Labour exchange team
Tổ hợp công nghiệp	Agro-industrial unit
Vùng kinh tế mới	New economic zones

Note on province names

Since 1975, the names of provinces in southern Vietnam have changed a number of times. In particular, between 1976 and the 1980s, many provinces were amalgamated to form larger ones. From 1990, the reverse process occurred, and some large provinces were divided up. The following list gives the names of the provinces in the Central Coast region and in the Mekong Delta in the period 1976–90 and the names used since 2005.

Provinces in the Mekong Delta

1976–90	2005
1. Long An	1. Long An
2. Tiền Giang	2. Tiền Giang
3. Đồng Tháp	3. Đồng Tháp
4. An Giang	4. An Giang
5. Kiên Giang	5. Kiên Giang
6. Minh Hải	6. Cà Mau
7. Hậu Giang	7. Bạc Liêu
8. Cửu Long	8. Sóc Trăng
9. Bến Tre	9. Hậu Giang
	10. Cần Thơ City
	11. Vĩnh Long
	12. Trà Vinh
	13. Bến Tre

Provinces in the Central Coast region

1976–90	2005
1. Bình Trị Thiên	1. Quảng Bình
2. Quảng Nam-Đà Nẵng	2. Quảng Trị
3. Nghĩa Bình	3. Thừa Thiên Huế
4. Phú Khánh	4. Quảng Nam
5. Thuận Hải	5. Đà Nẵng
	6. Quảng Ngãi
	7. Bình Định
	8. Phú Yên
	9. Khánh Hòa
	10. Ninh Thuận
	11. Bình Thuận

Note on measurements

The measurement of land areas and paddy yields varies across regions of Vietnam in terms of the unit used and its value. Villagers in Quảng Nam use *sào* (equal to 500 square metres), *thước* (one-fifteenth of 1 *sào*), *mẫu* (equal to 10 *sào* or 0.5 hectare) and hectares (equal to 20 *sào*) to measure their land area. They often use *ang* (equal to 4.5 kilograms of paddy) to measure their paddy yield. Meanwhile, more productive villages in An Giang use *công* (1,000 sq m), *mẫu* (10 công or 1 ha) and hectares to measure their land area. Villagers often use *gia* (equal to 20 kg) to measure their paddy yield.

1

Introduction

Vietnam has long been primarily an agrarian country. Land has always been an essential source of livelihoods, security and social status for the peasantry. Land is not only an important means of production, but also an important means of wealth, and has historically provided the strongest base for social and political power.[1] In other words, land is the major concern not only for peasants, but also for political leaders competing for power and people's allegiance and support. As in many other agrarian countries, in Vietnam, agrarian reforms have been carried out intermittently throughout its history, aimed at either stabilising existing power structures or consolidating new ones.

Soon after Vietnam was reunified, leaders of the Vietnamese Communist Party (VCP) in the central government launched full-scale social, economic and political reforms in the south to bring it into line with conditions in the socialist north and reunify the country politically, socially and economically. VCP leaders called this reform scheme the 'socialist revolution', and its aim was to transform Vietnam into a socialist country similar to other socialist states. The socialist revolution included socialist transformation and building. Socialist transformation was aimed at converting non-socialist elements into socialist ones, replacing private ownership of the main means of production with public ownership (collective and state) and eliminating institutions perceived to be 'old' and 'backward' and responsible for

---

1   Christodoulou, D. (1990), *The Unpromised Land: Agrarian Reform and Conflict Worldwide*, London: Zed Books, p. 22.

class exploitation in order to build 'new and advanced' ones. Socialist building meant establishing and reinforcing new (socialist) production relations, new productive forces and institutions and a new culture.

The VCP's leaders reasoned that, through socialist tools of planning and 'proletariat dictatorship', they could successfully build up large-scale socialist production and a 'rationally' structured agro-industrial economy, the two equally balanced legs of which (agriculture and heavy industry) would help move the economy rapidly forward. The party leaders also believed that, under socialist production relations and systems of ownership, Vietnam would be able to end poverty and class exploitation and become an advanced country with a socialist system of large-scale production.[2]

Having won the war and achieved the political reunification of Vietnam, the VCP strongly believed it could succeed in carrying forward the socialist revolution by building a centrally planned economy with large-scale production—a task it had not accomplished in the north. As Vietnam was an agricultural country, the VCP leaders considered agrarian reform a key component of the socialist revolution. Their reform or 'socialist transformation of agriculture and agricultural collectivisation' in the south had two main components: land redistribution and collectivisation. Redistribution was considered an important initial step of socialist agrarian reform. Socialist large-scale production or collective farming was the end goal of the transformation project. With high expectations of their capacity, the VCP leaders believed they could complete these projects within a few years.

This book shows that the results of land redistribution and agricultural collectivisation in the south varied from region to region. It also shows that, overall, socialist agrarian reform fell short of leaders' visions and expectations. There are two main reasons for this: regional differences and local politics. In the Central Coast region, the initial conditions seemed to be favourable for collective farming. Prolonged and destructive war had rendered most peasant households poor, and the social and economic structures of rural communities were flattened and relatively homogeneous. The region's new local authorities were

---

2   Đảng Cộng Sản Việt Nam [hereinafter ĐCSVN] (2004), Nghị quyết lần thứ 24 của Ban chấp hành Trung ương Đảng khóa III [Resolution No. 24 of the Third Party Central Committee], in ĐCSVN, *Văn Kiện Đảng Toàn Tập: Tập 36, 1975* [*Party Document: Volume 36, 1975*], Hà Nội: NXB Chính Trị Quốc Gia, p. 383.

quickly consolidated thanks to a considerable number of returned southerners and ex-revolutionaries who were able to fill government positions. They were familiar with and loyal to VCP policies and were able to mobilise people to implement these.

In contrast, in the Mekong Delta, conditions even initially were unfavourable for collective farming. Living in more favourable environmental and socioeconomic conditions and with less devastation from the war, peasant households in the Mekong Delta were better off than their Central Coast counterparts. Their social structure was also highly stratified and diverse: land reforms before reunification had led to the development of a middle strata of peasants and, by 1975, this group accounted for a majority of the peasantry and was largely engaged in commercial agriculture. They preferred individual to collective farming. In addition, the new local authorities in the Mekong Delta and elsewhere in the Southern Region (*Nam Bộ*) found it difficult to exert social control and carry out the VCP's postwar policies. Due to a shortage of cadres to fill new positions, local authorities had to recruit new cadres, a majority of whom were not former revolutionaries or southerners returning from the north. They were not familiar with the VCP leaders' post-1975 land redistribution and collectivisation policies and showed little enthusiasm for them.

Thus, implementing land redistribution and collectivisation was completed faster and more easily in the Central Coast than in the Mekong Delta region. By 1980, the Central Coast had largely completed its socialist agricultural transformation. Meanwhile, the Mekong Delta failed to meet its target and, by the end of 1980, only 8 per cent of peasant households and 6 per cent of agricultural land had been brought under collective farming. The slower transformation in the Mekong Delta was a result not only of stronger peasant resistance, but also of local cadres' lack of commitment to the socialist project. Opposition to collectivisation and land redistribution came from both landowners and the intended beneficiaries of these reforms, some of whom even engaged in open and confrontational resistance and other kinds of politics. Meanwhile, local cadres were unenthusiastic about and lax in implementing the socialist transformation policy; they took steps to implement it only when higher-level authorities pressed them to do so.

However, being resolute and persistent in their efforts to build socialist large-scale production, VCP leaders launched numerous directives and campaigns to urge local authorities in the south to complete the process. Only in the mid-1980s, after a decade of great effort, struggle and several policy modifications to ease local resistance, did authorities in the Mekong Delta and elsewhere in the Southern Region announce the completion of agrarian reform and collectivisation. Despite their efforts to establish collectives and bring peasants into these structures, VCP leaders were unable to direct peasants and local cadres to behave in line with their expectations. Thus, collective farming in Vietnam's southernmost region performed poorly and failed substantially to achieve its stated goals.

Although villagers in the Central Coast appeared to comply with the policy of collectivisation, they tried their best to maximise their individual earnings regardless of the outcomes of collective work. They undertook collective work carelessly and deceitfully, and often stole collective inputs and equipment and encroached on collective land, while they devoted time and material investments to their own household's economic activities, often at the expense of collective farming. In the Mekong Delta, many tried to evade collective farming as much as possible. Some joined collectives but did not actually participate in collective work; some participated in the work but just went through the motions; most did not take care of collective property. Many spent most of their time and effort on making a living somewhere else.

Local cadres in both places also manifested various forms of misbehaviour. They were caught between their orders from the top and the reality of the peasants they governed. With a lack of pressure from higher authorities, cadres, especially in the Mekong Delta, were reluctant to enforce the socialist transformation policy; often they modified policies to accommodate villagers' concerns and to protect local interests. However, when under pressure from the central government, local cadres carried out policies hastily and modified them to make them easier to implement, with little regard for either the overall purpose of the state's policies or villagers' interests. In addition, several local cadres increasingly abused their positions and became self-serving. They managed collectives poorly, embezzled a considerable amount of agricultural inputs and produce and misappropriated peasant land.

Despite numerous campaigns by the central and provincial authorities to crack down on and correct such 'bad behaviour', these problems did not disappear but seemed to increase over time.

As in the north, in both the Central Coast and the Mekong Delta, due to evasion and noncompliance by peasants and local cadres, collective farming performed poorly. Although collectivisation helped improve irrigation systems, increased the number of crops per year and succeeded in introducing new seeds and technology, it performed poorly compared with the family-based farming that it tried to eliminate and replace. In fact, collective farming could not produce sufficient food for the society. By the mid-1980s, Vietnam faced a serious fall in food production and was on the brink of an economic crisis. In this context, the Sixth National Congress (in December 1986) released a 'renovation policy' (*chính sách đổi mới*) that abandoned the centrally planned economy and adopted a market-based one.

The economic component of *đổi mới* opened the way to new forms of ownership and management, and the resurgence of the private sector and the market. In this context, collective farming faced even more difficulties. Local cadres became even more lax about management and abused their positions to make use of market opportunities for personal benefit at the expense of collective farming. Meanwhile, many villagers refused or were not able to pay their debts or fulfil their obligations to the collectives. Some even returned land or abandoned it when they saw that their contracted land was unprofitable.

The ultimate consequences of such deviant practices were a gradual demise in the efficiency of collective farming and consistent falls in food production and peasants' living standards. Faced with local food shortages, villagers and local cadres had to initiate new farming arrangements. The aim of these experiments was to encourage villagers to work on collective land and keep collectives alive; however, this gradually derailed collective farming from its original intention and amounted to an informal return to individual farming. In other words, the failure of socialist agrarian reforms and collectivisation in the postwar era resulted significantly from the widespread involvement by peasants and local officials in everyday practices that deviated from official guidelines and the VCP leaders' expectations.

A few studies have addressed the agrarian reforms in southern Vietnam from 1975 to the late 1980s. Most are short articles, but they also include a few books and dissertations in Vietnamese.[3] Recent books by Đặng Phong and Huy Đức reveal some secrets about how political and economic decisions were negotiated and formulated at the top leadership level.[4] However, the existing literature provides scant detail and insufficient analysis of VCP leaders' approach to post-1975 agrarian reform and how such policies were carried out at the local level. In addition, the existing literature is largely silent on explanations of variations in villagers' behaviour and policy outcomes across regions within southern Vietnam, and on the key factors contributing to the failure of socialist large-scale production and the shift in state agrarian policies.

Writing on northern Vietnam's agrarian reform, Ben Kerkvliet gives a rich account of everyday politics and convincingly explains how it significantly contributed to the demise of collective farming and modifications to Vietnam's national policies. According to Kerkvliet, everyday politics includes 'quiet, mundane, and subtle expressions and acts that indirectly and for [the] most part privately endorse, modify, or resist prevailing procedures, rules, regulations, or order'.[5] It involves

---

3   Đào Duy Huấn (1988), Củng cố và hoàn thiện quan hệ sản xuất xã hội chủ nghĩa trong nông nghiệp tập thể hiện nay ở vùng Đồng Bằng Sông Cửu Long [Solidifying and perfecting socialist production relations in the agriculture of the Mekong Delta], PhD thesis, Học Viện Nguyễn Ái Quốc, Hà Nội; Ngo Vinh Long (1988), Some aspects of cooperativization in the Mekong Delta, in D. Marr and C. White (eds), *Postwar Vietnam: Dilemmas in Socialist Development*, Ithaca, NY: Cornell University Press; Huỳnh Thị Gấm (1998), Những biến đổi kinh tế xã hội ở nông thôn Đồng bằng sông Cửu Long 1975–1995 [Socioeconomic changes in the Mekong Delta from 1975–1995], PhD thesis, Đại Học Khoa Học Xã Hội Nhân Văn, Hồ Chí Minh; Nguyễn Sinh Cúc (1991), *Thực Trạng Nông Nghiệp, Nông Thôn và Nông Dân Việt Nam 1976–1990* [*Agricultural and Rural Development in Vietnam 1976–1990*], Hà Nội: NXB Thống Kê; Trần Hữu Đính (1994), *Quá trình biến đổi về chế độ sở hữu và cơ cấu giai cấp nông thôn Đồng Bằng Sông Cửu Long (1969–1975)* [*The Process of Ownership and Class Structure Change in Rural Mekong Delta, 1969–1975*], Hà Nội: NXB Khoa Học Xã Hội; Lê Thị Lộc Mai (2001), Quá trình giải quyết vấn đề ruộng đất và phát triển nông thôn ở Vĩnh Long giai đoạn Đổi mới 1986–1996 [Dealing with land problems to facilitate rural development in Vĩnh Long in the period 1986–1996], Masters thesis, Đại học Khoa học Xã hội & Nhân văn, Hồ Chí Minh; Nguyễn Minh Nhị (2004), *An Giang: Lịch sử tháo gỡ đột phá và chủ động hội nhập kinh tế thế giới* [*An Giang: The History of Breakthroughs and Active Integration into the World Economy*], 15 August, Long Xuyên: Sở Nông Nghiệp và Phát Triển Nông Thôn An Giang; Hicks, N. (2005), Organizational adventures in district government, PhD thesis, The Australian National University, Canberra.
4   Đặng Phong (2009), *Tư duy kinh tế Việt Nam 1975–1989* [*The Economics of Vietnam 1975–1989*], Hà Nội: Nhà Xuất Bản Trí Thức; Huy Đức (2012), *Bên Thắng Cuộc* [*The Winning Side*], 2 vols, Giai Phong: OsinBook.
5   Kerkvliet, B. J. (2005), *The Power of Everyday Politics: How Vietnamese Peasants Transformed National Policy*, Ithaca, NY: Cornell University Press, p. 22.

little or no organisation. Due to their discontent with collective farming, peasants relied on their own strategies for survival—aimed at making a living, raising their families and wrestling with daily problems. For example, they tried their best to 'minimize the cooperative's claim on their labour and to maximize their household-based production'.[6] They tried to minimise their effort, time and costs while maximising work-points. They took advantage of any opportunity to steal collective inputs, produce, time and equipment, while devoting time, materials and effort to their own household plots. According to Kerkvliet, although these tactics were low-key, dispersed, largely unorganised and nonconfrontational, and were often carried out individually, they occurred in many places at the same time and the cumulative effects of thousands of such actions therefore had a huge impact on the performance of collective farming in Vietnam.

James Scott also claims the outcome of state policies almost always 'depends on the response and co-operation of real human subjects'. In innumerable instances, ordinary people have played significant political roles. In particular, they have proven capable of undermining, resisting or even transforming state policies, even in authoritarian settings. For example, collectivisation of Soviet agriculture and 'villagisation' in Tanzania failed badly, largely because they encountered strong resistance from peasants, 'including flight, unofficial production, and trade, smuggling, and foot dragging'.[7] The goal of such resistance is to thwart material extraction by states or dominant classes rather than to directly overthrow or transform them. Scott has shown that, in socialist states, foot-dragging and evasion in response to unpopular forms of collective agriculture can short-circuit grandiose policies dreamed up by national leaders.

In this book, I adopt and expand on James Scott's notion of everyday forms of resistance and Ben Kerkvliet's concept of everyday politics to examine how everyday politics played out under and affected post-1975 national agrarian policies in different regions of southern Vietnam. By focusing on two case studies—Quảng Nam-Đà Nẵng (QN-ĐN) province (now two separate provinces, Quảng Nam and Đà Nẵng) in the Central Coast region and An Giang province in the Mekong Delta—

---

6   Ibid., p. 2.
7   Scott, J. C. (1998), *Seeing Like a State: How Certain Schemes to Improve the Human Condition Have Failed*, New Haven, CT: Yale University Press, p. 247.

the book traces the rationale for and content of post-1975 agrarian reforms and socialist building in these two places. It examines how national agrarian policies were carried out and shows how everyday politics at the local level was able to divert the direction of policies issued by top-level VCP leaders. In particular, it examines similarities and differences in peasants' and local cadres' behaviours and politics in these two places, and the effects of local conditions and local politics on the ability of local authorities to implement the post-1975 agrarian reforms. As such, it provides regionally specific insights into postwar experiences of socialist agrarian reform and the local factors that led to the failure of and a shift in such policies nationwide.

This book argues that peasants' everyday politics and local cadres' malpractices and corruption played an important role in derailing the VCP's post-1975 agrarian reforms in both the Central Coast and the Mekong Delta. Significant variation in the outcome of land reform and collectivisation policies arose from regional differences with regards to socioeconomic conditions and political capacity, and the forms and magnitude of peasant politics and local cadres' noncompliance. Despite the variation, as in the north, in the Central Coast and the Mekong Delta, collective farming performed poorly; it was inferior to the private farming that it tried to replace and could not produce sufficient food for the society. Thus, collective farming failed to fulfil the VCP's objectives of increased productivity, improved living standards for peasants and ending class exploitation in the countryside. To cope with production problems, local cadres and their fellow villagers had to modify or initiate new farming arrangements. Despite VCP leaders' persistent efforts to strengthen and consolidate collective farming and crack down on local evasion, resistance and malpractice, these problems could not be rooted out, but rather persistently increased over time. Faced with food shortages and many other problems related to collective farming, VCP leaders gradually modified their policies, and eventually accepted and officially endorsed local initiatives. The accumulated effects of piecemeal policy modifications led to collective farming gradually departing from the VCP's original intentions for it. Finally, national leaders shifted their policies back to household-based farming in the late 1980s and, from then, collectives lost their purpose and were gradually dismantled nationwide.

Approach and methods

For my research, I selected the Mekong Delta in the south and the Central Coast in the north of what I call southern Vietnam, the territory formerly under the governance of the Republic of Vietnam (1955–75). In both regions, food crops, especially rice, have long been the primary farm produce. This sets the two regions apart from the South-East Region and the Central Highlands, in which industrial crops, rather than food crops, have been prominent—rubber and fruit trees in the former and coffee and rubber in the latter. While the Mekong Delta and the Central Coast have similarities, there are also notable differences. Population density on arable land in the Mekong Delta is lower than in the Central Coast. Village settlements in the Mekong Delta follow rivers and channels whereas most villages in the Central Coast are relatively isolated and surrounded by hedges and rice fields. Pre-1975 agrarian reforms seem to have had a greater impact in the Mekong Delta than in the Central Coast. This led to more commercial farming of food (including rice) and greater complexity in rural social structures in the Mekong Delta compared with the Central Coast, which had greater levels of subsistence agriculture and relatively homogeneous rural communities before 1975.

Within the two selected regions, I focused on two provinces: Quảng Nam province (previously part of Quảng Nam-Đà Nẵng) in the Central Coast and An Giang province in the Mekong Delta. Within each province, I focused on one district: Thăng Bình in Quảng Nam and Chợ Mới in An Giang. In both districts, agrarian reform and collectivisation campaigns after 1975 were rather intense, perhaps more than in some other parts of the two provinces. In Thăng Bình district, I examined two communes (*xã*): Bình Lãnh, where provincial and district authorities established a pilot collective on 30 October 1977, and Bình Định, which underwent normal collectivisation. Most of my interviews were in Bình Lãnh's Hiền Lộc village and Bình Định's Thanh Yên village. In Chợ Mới, An Giang, I focused on Long Điền B commune in which provincial and district authorities built pilot production units (*tập đoàn sản xuất*) in the summer–autumn crop season of 1979 (after failing to experiment with collectives in the province). In Long Điền B commune, I interviewed villagers across its eight different hamlets.

Interviews were one main source of data. I conducted two rounds of fieldwork: from September to October 2004 and from June to December 2005. I spent most of my fieldwork time interviewing ordinary villagers and current and former officials at different levels who had experienced the socialist transformation of agriculture and collective farming from 1975 to the late 1980s. Most of these people were more than 50 years old. The interviews were open-ended, rather than structured questionnaires. I asked people about their experiences, observations and their assessments of post-1975 agrarian issues related to my broad research questions. The specific questions asked of each informant varied depending on the person's background and involvement, their comments and the information they provided, and what I had learned during the course of my research. When I found it convenient, I asked permission to tape-record interviews.

In Hồ Chí Minh City, I was able to interview two officials who had previously been staff members of the Committee for Southern Agricultural Transformation (Ban Cải Tạo Nông Nghiệp Miền Nam, or BCTNNMN). In An Giang, I was able to interview three former staff members of the An Giang Committee for Agricultural Transformation (Ban Cải Tạo Nông Nghiệp An Giang, or BCTNNAG). In Chợ Mới, I interviewed three district officials and more than 15 commune, hamlet and production unit cadres who were directly engaged in carrying out agrarian policies from 1975 to the late 1980s. In Quảng Nam, I was able to interview three provincial officials, one local journalist, two district officials and more than 15 commune, collective and brigade cadres.

In terms of ordinary villagers, I was able to stay in selected villages for a total of four months in each province, so I had many opportunities to chat with and interview individuals and groups. In particular, I was able to interview more than 100 male and female villagers in each province. The interviews were carried out mostly in their homes, varying in length from 20 minutes to two hours. Some individuals were interviewed more than once. In my first round of fieldwork, I took notes to record my interviews. However, in the final round of fieldwork, thanks to the rapport established, I was able to tape-record more than 60 interviews in each province. For the safety of informants, I generally use pseudonyms when referring to them.

Documents are another primary source. In Hồ Chí Minh City, I was able to access and photocopy some relevant books, dissertations and national newspapers such as *Nhân Dân* (*The People*), *Đại Đoàn Kết* (*Great Unity*) and *Sài Gòn Giải Phóng* (*Saigon Liberation*) in the General Sciences Library (Thư viện Khoa Học Tổng Hợp) and Social Sciences Library (Thư viện Khoa Học Xã Hội). When interviewing staff members of the BCTNNMN, I was given some valuable committee reports. In An Giang and Quảng Nam, I acquired relevant materials— published and unpublished—from numerous government agencies at different levels, such as people's committees, departments of agriculture and rural development, departments of statistics and libraries. These documents include reports, surveys, statistics, historical records and studies done by commune, district, provincial and national agencies.

Importantly, I was able to access and copy local newspapers, magazines and reports, ranging from 1974 to the early 1990s in the general library (Thư Viện Tổng Hợp) in Đà Nẵng City for Quảng Nam-Đà Nẵng (QN-ĐN) province and from 1978 to the early 1990s in the An Giang Library in Long Xuyên City, An Giang province. (Unfortunately, An Giang newspapers from before 1979 were unavailable because they were destroyed in a flood in 1978.) Although newspapers were organs of the VCP with a political propaganda purpose, I found them to be a valuable source of information if read carefully and selectively. The newspapers covered a wide range of information on national and provincial policies, statistics, policy implementation and results and daily struggles at the village level across different places in each province. It was common for articles to reveal or criticise policy shortcomings and problematic activities that had occurred a few years earlier as well as providing more accurate statistics some years after the event than had been available at the time. The newspapers often carried debates over agrarian issues, and local papers also frequently published readers' letters or petitions regarding their land, property and other agrarian issues or their complaints about the corrupt practices of local cadres.

Organisation

Following this introduction to the book's themes and approach, Chapter 2 examines VCP leaders' objectives for the post-1975 agrarian reforms. It also reviews the pre-1975 agrarian reforms and points out how these resulted in regional differences within southern

Vietnam. I found there were numerous political, social and economic objectives for the post-1975 agrarian reforms. The primary objective was consolidating power and building socialism, but VCP leaders also hoped reforms would solve their postwar economic problems and modernise the south's agriculture.

Chapter 3 looks at post-1975 land reform and other preparations for collectivisation in QN-ĐN in the Central Coast and An Giang province in the Mekong Delta. I found that, in QN-ĐN, the local authorities quickly consolidated their power and successfully carried out preparatory policies such as land restoration and reform. Meanwhile, authorities in An Giang and many other provinces in the Mekong Delta faced difficulties consolidating power and had problems implementing preparatory policies such as land reform and crop conversion.

Chapter 4 examines the building of pilot collective organisations and the acceleration of collectivisation. I found that collectivisation in QN-ĐN was achieved rapidly, but it faced major difficulties in An Giang. The reason was that collectivisation faced weaker peasant resistance in QN-ĐN than in An Giang. Moreover, local cadres in QN-ĐN were more loyal to the socialist transformation policy than their counterparts in An Giang.

Chapter 5 examines the performance of collective organisations under the work-points system (1978–81). I found peasants' everyday politics and local cadres' malpractices contributed significantly to the poor performance of collective farming. As the process of collectivisation continued, peasants in QN-ĐN chased work-points at the expense of collective farming. Meanwhile, many of their counterparts in An Giang were not even undertaking collective work; they tried their best to evade or abandoned collective work as much as possible. And cadres in both places managed collectives poorly. Despite many cadres in QN-ĐN being loyal to the VCP's agrarian policies, several abused their power at the expense of peasants' and collectives' interests. Meanwhile, cadres in An Giang were unenthusiastic about collective farming and exercised slack management of labour, finance, production and distribution of produce. Several also abused their positions and became self-serving. Despite numerous campaigns by the authorities to strengthen collective farming in both places, the malpractice among local peasants and local cadres could not be reduced.

Chapter 6 examines the modifications of the VCP's agrarian policies and the adoption of the product contract system in An Giang and QN-ĐN. It also examines the second wave of land reform and collectivisation and the strengthening of collective farming from 1981 to the late 1980s in An Giang and elsewhere in the Southern Region. The chapter shows that, in the first few years after the adoption of product contracts, the performance of collective farming improved significantly, in both QN-ĐN and An Giang provinces. In An Giang, product contracts saved some production units from collapse and facilitated the completion of the second wave of land redistribution and collectivisation because peasants seemed to accept the product contract system more than the work-points system. However, from 1985 to the late 1980s, collective farming in both An Giang and QN-ĐN was in crisis and similar problems arose in both places, owing to the effects of local politics. Performance again declined, peasants' living standards dropped alarmingly and a new class of exploiters started to emerge.

Chapter 7 examines in depth the relationship between local politics, the performance of collective farming under product contracts (1981 to the late 1980s) and decollectivisation initiated at the local level. I found that collective farming under the product contract system continued to face major shortcomings and the impact of local politics. Despite differences in magnitude, peasants and local cadres in both places manifested similar forms of politics and noncompliant behaviour. Villagers in the Central Coast tried their best to enlarge their household economies by capturing collective resources, land and labour at the expense of the collective economy. Meanwhile, An Giang villagers tried their best to ensure their livelihoods by undertaking wage work elsewhere and using collective resources for their daily needs rather than investing in their collective fields. In the late 1980s, villagers in both places increasingly tried to avoid repaying debts and fulfilling their obligations to the collective; they wanted to return collective land or abandoned it when they saw that collective farming was unprofitable.

Local cadres in both places increasingly abused their power and became self-serving. QN-ĐN cadres shifted most collective work tasks to peasants and did not perform their own services well. They stole collective property and irrationally increased the quotas and agricultural input prices for peasants. Meanwhile, An Giang cadres assigned almost all farming tasks to peasants and increasingly embezzled resources, misappropriated peasant land, mismanaged collective funds, monopolised agricultural services and—in the words of my interviewees and official reports alike—oppressed the masses. It was common for local cadres to misappropriate land anywhere they could and they had more land than ordinary people.

As discussed in Chapter 7 and the book's conclusion, these kinds of local politics had a huge adverse effect on the survival of collective organisations and led to the failure of collective farming. In the late 1980s, local authorities, especially in An Giang, started to rethink the direction and purpose of collective farming and created policies to correct previous shortcomings. New individual farming arrangements beyond the official orthodoxy were adopted and became widespread, not only in the Mekong Delta, but also in the Central Coast and elsewhere in Vietnam. At first, VCP leaders put great effort into cracking down on such practices, but they were unable to control them and gradually recognised their benefits and accepted them. Finally, they endorsed local initiatives by shifting their policies back to household-based farming.

Map 1.1 Southern Vietnam, 2005

Source: Nhà xuất bản Bản Đồ (2005), *Vietnam's Administrative Atlas*, Hà Nội: NXB Bản Đồ, p. 8.

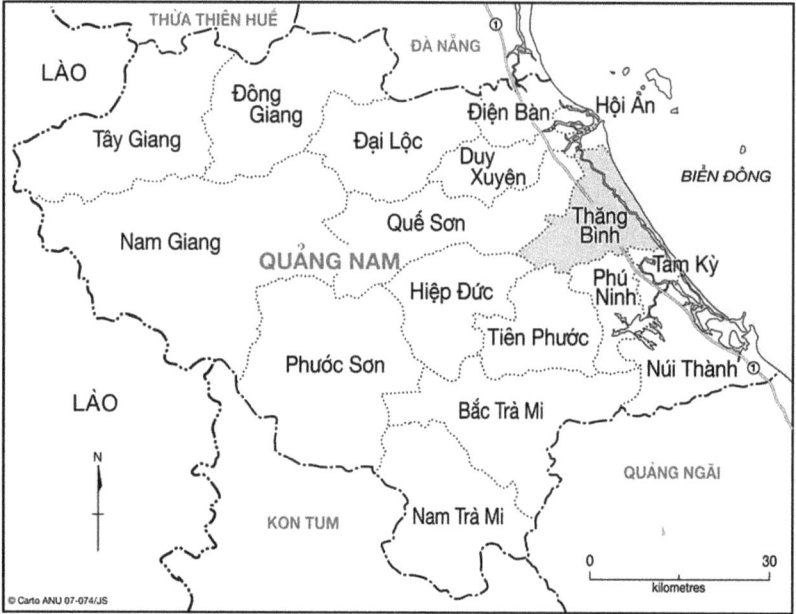

Map 1.2 Quảng Nam administrative map, 2005

Source: Nhà xuất bản Bản Đồ (2005), *Vietnam's Administrative Atlas*, Hà Nội: NXB Bản Đồ, p. 48.

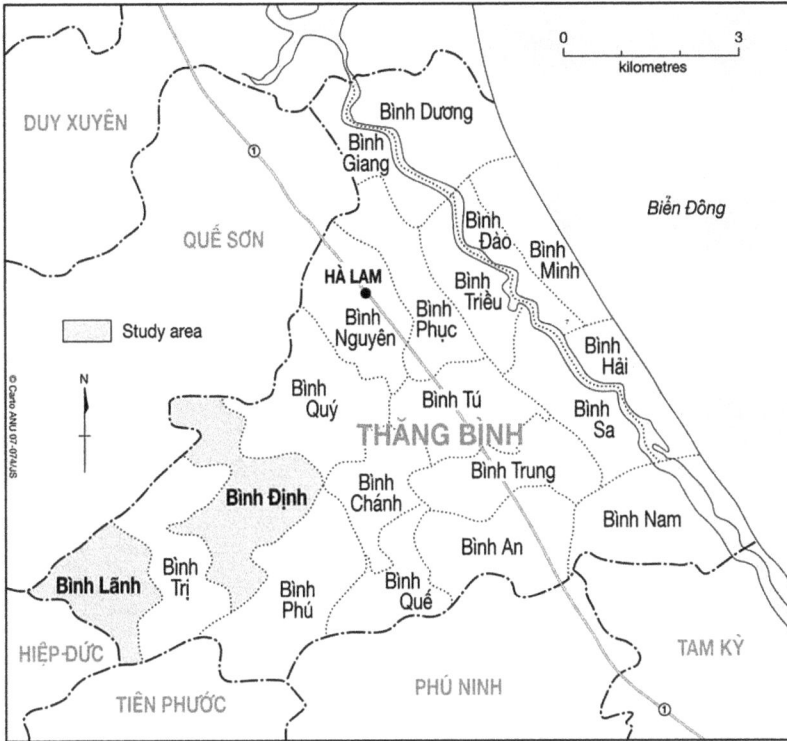

Map 1.3 Thăng Bình district map, 2005

Source: Nhà xuất bản Bản Đồ (2005), *Vietnam's Administrative Atlas*, Hà Nội: NXB Bản Đồ, p. 48.

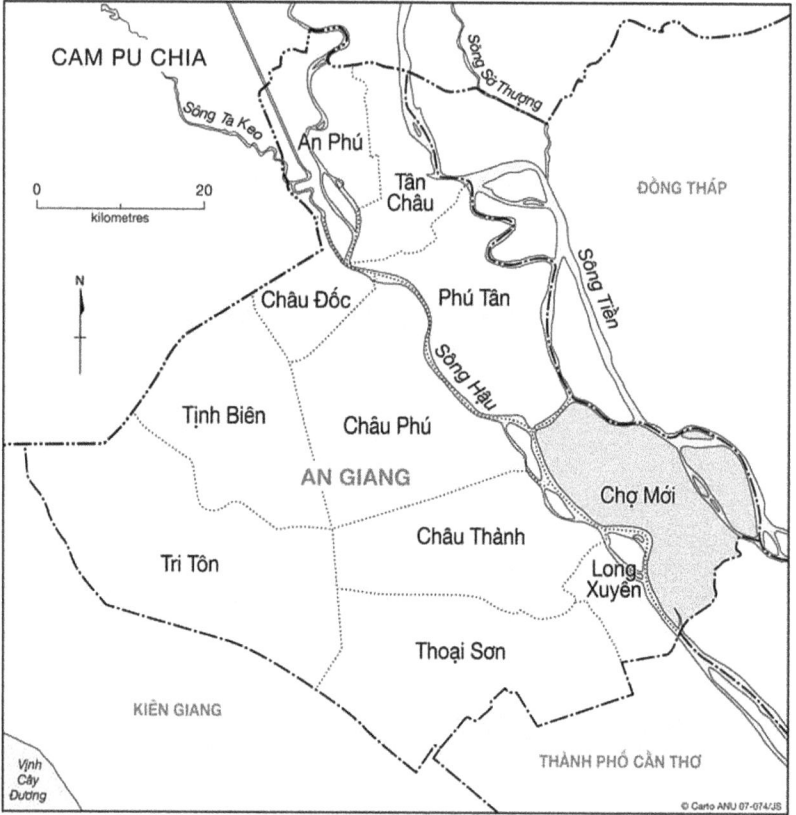

Map 1.4 An Giang administrative map, 2005

Source: Nhà xuất bản Bản Đồ (2005), *Vietnam's Administrative Atlas*, Hà Nội: NXB Bản Đồ, p. 73.

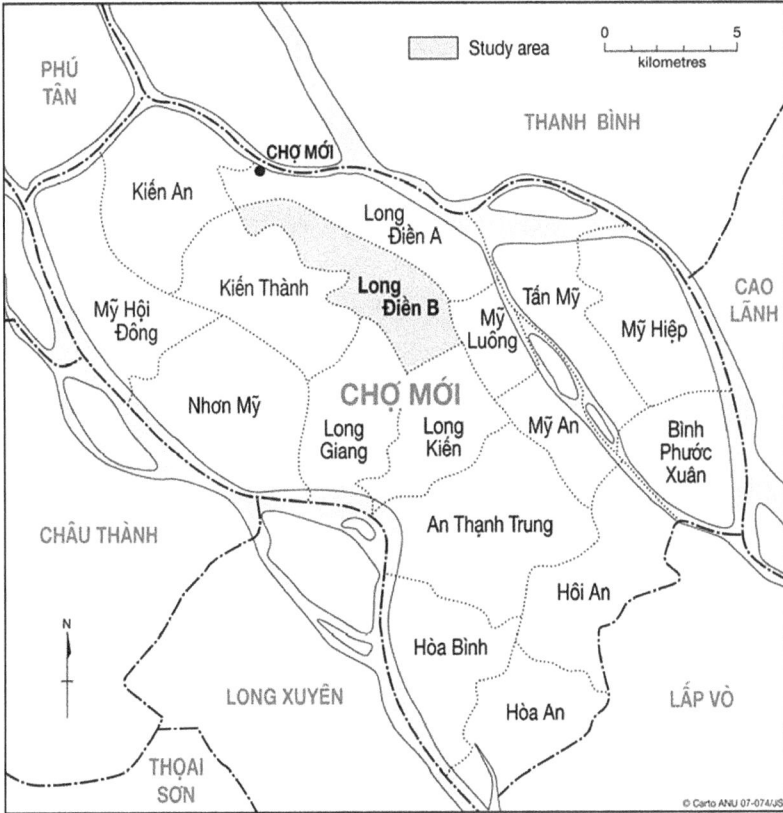

Map 1.5 Chợ Mới district map, 2005

Source: Nhà xuất bản Bản Đồ (2005), *Vietnam's Administrative Atlas*, Hà Nội: NXB Bản Đồ, p. 73.

2

Vietnamese Communist Party leaders' reasons and objectives for post-1975 agrarian reform

Introduction

After the military victory of April 1975, southern Vietnam was under the control of Hanoi's government in general and the Vietnamese Communist Party (VCP) leaders in particular. The VCP very quickly decided to remake the south in line with the socialist north. They called for a 'socialist revolution' or 'socialist transformation and building' in the south, a key component of which was the socialist transformation of agriculture.

According to Đặng Phong, the Secretary-General of the VCP, Lê Duẩn was the principal architect of North Vietnam's economic model during 1960–75 and unified Vietnam's economic model from 1975 until his death in 1986. Therefore, Vietnam's post-1975 economic model in the south was heavily influenced by Lê Duẩn's thoughts and North Vietnam's model.[1] Postwar 'economic problems and the reunification of Vietnam after so many years of war', Christine White contends, created 'an unfavourable context for an open, innovative, and creative approach to experimentation with alternative routes to socialism'.[2]

---

1 Đặng Phong, *The Economics of Vietnam 1975–1989*, pp. 72–3.
2 White, C. P. (1988), Alternative approaches to the socialist transformation of agriculture in postwar Vietnam, in D. Marr and C. White (eds), *Postwar Vietnam: Dilemmas in Socialist Development*, Ithaca, NY: Cornell University Press, pp. 134–43.

This chapter will not focus on internal debates and decision-making processes at the top level. Instead, it examines the main reasons the VCP decided to carry out this process in the south. It also explores the development model the VCP pursued and its rationale. In particular, it examines the original objectives, content and steps of the post-1975 socialist agricultural transformation and socialist building. Before examining these, however, it is essential to provide an overview of pre-1975 land and agrarian reforms and their legacy.

Overview of southern Vietnam's pre-1975 land tenure and agrarian reforms

In precolonial times, Vietnam had three intertwined and competing forms of landownership: state, village and individual.[3] Large portions of village land were communally owned, inalienable by law and periodically distributed among the male inhabitants. Those who were outsiders or not born in the village were excluded from a share of communal land.[4] From the seventeenth to the nineteenth century, Vietnamese territory gradually expanded to the south and reached the vast plains of the Mekong Delta. To encourage this southward land reclamation, the state allowed peasants to claim and own as much land as they wanted. This led to a land tenure system in the Southern Region that was different to that in the rest of Vietnam. Private land became dominant while levels of communal land were insignificant.[5]

French colonial rule brought about a major upheaval in Vietnam's land tenure system. French policies favoured large landowners at the expense of traditional small landowners. The French government granted large tracts of land—whether free, at cheap prices or at auction—to French colonists and their Vietnamese collaborators.[6] By the 1930s, all regions

---

3   Vũ Huy Thúc (1979), *Tìm hiểu chế độ ruộng đất Việt Nam nửa đầu thế kỷ XIX* [*Examining Vietnam's Land Tenure System in the First Half of the Nineteenth Century*], Hà Nội: NXB Khoa Học Xã Hội, p. 11.
4   White, C. (1981), Agrarian reform and national liberation in the Vietnamese revolution 1920–1957, PhD thesis, Cornell University, Ithaca, NY, p. 28.
5   Nguyễn Đình Đầu (1992), *Chế Độ Công Điền Công Thổ Trong Lịch Sử Khẩn Hoang Lập Ấp ở Nam Kỳ Lục Tỉnh* [*Land Tenure System in the Southern Region of Vietnam in the History of Land Reclamation*], Hồ Chí Minh: NXB Trẻ, pp. 61, 82; Vũ Huy Thúc, *Examining Vietnam's Land Tenure System*, pp. 215–16.
6   White, Agrarian reform and national liberation, pp. 31–2.

of Vietnam faced a severely unequal pattern of landholding distribution. In the Mekong Delta, a small group of landlords owned much of the land and tenants farmed 80 per cent of cultivated land.[7] In 1953, near the end of colonial rule, the Bảo Đại government put forth an agrarian policy to compete with the Việt Minh's agrarian reforms. The land reform decrees (*cải cách điền địa*) that were issued advocated rent reduction, security of land tenure and modest restrictions on the maximum size of holdings.[8] This reform was unsuccessful, however, because the Bảo Đại government made no serious effort at implementation.[9]

After the collapse of French colonialism in 1954, Vietnam was divided until its reunification in 1975. In the north, the Democratic Republic of Vietnam pursued land redistribution and agricultural collectivisation. Land reform was carried out vigorously and violently between 1953 and 1957, with fields redistributed more or less equally between all farming households.[10] Land reform was considered a necessary and immediate step towards collectivisation.[11] By 1960, North Vietnam had completed collectivisation.[12] In the south, Prime Minister Ngô Đình Diệm also considered land reform a top national policy when he came to power. His Ordinance No. 57 (22 October 1956) called for land redistribution by limiting maximum holdings to 100 hectares plus 15 hectares of ancestral (patrimonial) land (*ruộng hương hỏa*). Any holding in excess of that limit was subject to expropriation (*truất hữu*).[13] Ordinance No. 57 was supposed to affect 2,035 Vietnamese landlords (including 12 in

---

7   Callison, C. S. (1983), *Land-to-the-Tiller in the Mekong Delta: Economic, Social, and Political Effects of Land Reform in Four Villages of South Vietnam*, New York: University Press of America, p. 39.
8   Trần Phương (1968), *Cách mạng ruộng đất ở Việt Nam [Land Revolution in Vietnam]*, Hà Nội: NXB Khoa Học Xã Hội, pp. 247–8.
9   Prosterman, R. L. and Riedinger, J. M. (1987), *Land Reform and Democratic Development*, Baltimore: Johns Hopkins University Press, p. 120.
10   Wiegersma, N. (1988), *Vietnam: Peasant Land, Peasant Revolution—Patriarchy and Collectivity in the Rural Economy*, New York: St Martin's Press, p. 139.
11   Moise, E. E. (1976), Land reform and land reform errors in North Vietnam, *Pacific Affairs* 49(1), pp. 70–92.
12   Kerkvliet, B. (1999), Accelerating cooperatives in rural Vietnam, 1955–1961, in B. Dahm, V. J. H. Houben, M. Grossheim, K. W. Endres and A. Spitzenpfeil (eds), *Vietnamese Villages in Transition: Background and Consequences of Reform Policies in Rural Vietnam*, Passau, Germany: Department of Southeast Asian Studies, University of Passau, pp. 53–88.
13   Quang Truong (1987), Agricultural collectivization and rural development in Vietnam: A north/south study (1955–1985), PhD thesis, Vrije Universiteit te Amsterdam, Amsterdam, pp. 141–2; Trần Phương, *Land Revolution in Vietnam*, pp. 252–3; Callison, *Land-to-the-Tiller in the Mekong Delta*, p. 43.

the Central Coast) with holdings of more than 100 hectares, and 200 French landlords. In other words, 650,000 hectares of land were to be expropriated.[14] However, by the end of 1967, only 275,000 hectares had been redistributed, to 130,000 families, accounting for less than one-eighth of South Vietnam's cultivated land and one-tenth of its tenant farmers.[15]

Following the fall of the Ngô Đình Diệm government in November 1963 and amid growing insecurity and political instability, successive governments in the south made no further efforts towards land reform until Nguyễn Văn Thiệu's tenure in 1967. On 26 March 1970, backed by the United States, Nguyễn Văn Thiệu's government launched its Land to the Tiller (LTTT) program, in the hope of gaining peasant support against the growing National Liberation Front (NLF). The LTTT program was similar to distribution programs carried out years earlier in Taiwan, South Korea and Japan.[16] By February 1975, some 1,136,705 hectares—nearly half of the rice-growing land in the south—had been redistributed. Under the LTTT program, 77 per cent of tenants became landowners.[17] In addition to the Saigon government's land reforms, the NLF carried out rent reduction and land redistribution in areas under its control, which were known as 'liberated areas' (*vùng giải phóng*). NLF reforms also contributed significantly to the rise of middle landowners. By 1969, middle peasants made up about 51–87 per cent of the rural population in NLF-controlled areas.[18]

---

14   Trần Phương, *Land Revolution in Vietnam*, pp. 265–72.
15   Prosterman and Riedinger, *Land Reform and Democratic Development*, p. 126.
16   Wiegersma, *Vietnam*, p. 191.
17   Callison, *Land-to-the-Tiller in the Mekong Delta*, pp. 327–32.
18   Elliott, D. W. (2003), *The Vietnamese War: Revolution and Social Change in the Mekong Delta 1930–1975*, Armonk, NY: M. E. Sharpe, pp. 1, 451. According to David Elliott, to analyse the politics of land, the VCP developed several class categories for rural society, including landlords (*địa chủ*), petty bourgeoisie (*tiểu tư sản*), capitalists (*tư sản*), rich peasants (*phú nông*), middle peasants (*trung nông*), poor peasants (*bần nông*) and landless peasants (*cố nông*). Middle peasants were those who owned sufficient land, farm animals and tools to support their families (p. 459).

In summary, between 1945 and 1975, land tenure patterns in the south gradually changed as a result of reforms carried out by the Việt Minh, governments in Saigon and the NLF, and also as a result of the transformation of rural society during the war.[19] By 1975 in the Mekong Delta, 70 per cent of the rural population were middle peasants who owned 80 per cent of the cultivated land, 60 per cent of the total farm equipment and 90 per cent of draught animals.[20] Unlike the agrarian sector in North Vietnam, which, at the beginning of the land reforms was dominated by landlords, the agrarian sector in the Southern Region was dominated by middle peasants who engaged largely in commercial agriculture. This large cohort of middle peasants wanted to continue to farm their own land and sell their own crops. They proved resistant to the post-1975 land redistribution and collectivisation in the south.[21]

Meanwhile, in the Central Coast region, the significant factor transforming the land tenure system was war rather than any pre-1975 reforms. The war had disrupted or destroyed any positive effect of land reforms carried out by governments in Saigon or the NLF. It had caused a large proportion of rural people to live in enclosed camps and much of their land had been abandoned. After the war, many peasants returned home without capital, draught animals or farm tools. Thus, soon after reunification, the agricultural sector in the Mekong Delta had reached a higher level of economic development than that in the Central Coast. The social structure and rural economy in the Mekong Delta were more diverse than in the Central Coast. These regional disparities contributed to differences in peasants' behaviour and the results of post-1975 agrarian policies.

---

19   Grossheim, M. (1999), The impact of reforms on the agricultural sector in Vietnam: The land issue, in B. Dahm, V. J. H. Houben, M. Grossheim, K. W. Endres and A. Spitzenpfeil (eds), *Vietnamese Villages in Transition: Background and Consequences of Reform Policies in Rural Vietnam*, Passau, Germany: Department of Southeast Asian Studies, University of Passau, p. 97.
20   Ngo Vinh Long (1988), Some aspects of cooperativization in the Mekong Delta, in D. Marr and C. White (eds), *Postwar Vietnam: Dilemmas in Socialist Development*, Ithaca, NY: Cornell University Press, p. 169.
21   Elliott, *The Vietnamese War*, p. 4.

Post-1975 agrarian reform in the south

Soon after the war, at the 24th plenum of the Third Party Congress, in September 1975, the VCP began planning how to bring the south into line with the north politically, socially and economically and make the whole nation socialist.[22] At this meeting, the party released a resolution that outlined the 'ongoing tasks of Vietnam's revolution in the new age', one of which was to 'accomplish national reunification and take the whole nation fast, vigorously and firmly to socialism'.[23]

On 25 April 1976, the official political reunification of the country came about through a national election to establish a unified National Assembly.[24] At the first session of this new assembly, in June 1976, General Secretary Lê Duẩn clarified the tasks of economic reunification for both the north and the south of Vietnam:

> [T]he north must continue speeding up the task of building socialism and improving socialist production relations; the south must proceed simultaneously on the task of socialist transformation and the task of socialist building.[25]

The aim, according to party leaders, was to transform non-socialist elements into socialist ones, replace private ownership of the main means of production with public ownership (collective and state) and eliminate perceived 'old' and 'backward' institutions and build 'new and advanced' ones. Socialist building meant establishing new, socialist production relations, new productive forces, new super-infrastructure and a new culture.

---

22 The Third Central Committee was the committee formally chosen at the time of the third party congress in 1960.

23 ĐCSVN (2004), Nghị quyết hội nghị lần thứ 24 của Ban Chấp Hành Trung Ương ĐCSVN, số 247/NQ-TW (ngày 29 tháng 9 năm 1975) [Resolution No. 247/NQ-TW (29 September 1975)], in ĐCSVN, *Văn Kiện Đảng Toàn Tập: Tập 36, 1975 [Party Document: Volume 36, 1975]*, Hà Nội: NXB Chính Trị Quốc Gia, p. 369. I used the electronic versions of the VCP's documents (*Văn Kiện Đảng Toàn Tập*) downloaded from the party's website: www.vcp.org.vn (accessed 13 February 2006).

24 Phạm Văn Chiến (2003), *Lịch Sử Kinh Tế Việt Nam [History of the Vietnamese Economy]*, Hà Nội: NXB Đại Học Quốc Gia, p. 152.

25 Lê Duẩn (2004), Toàn dân đoàn kết xây dựng tổ quốc Việt Nam thống nhất, xã hội chủ nghĩa [Calling for the whole country's solidarity to build a socialist and unified country], in ĐCSVN, *Văn Kiện Đảng Toàn Tập: Tập 37, 1976 [Party Document: Volume 37, 1976]*, Hà Nội: NXB Chính Trị Quốc Gia, p. 140.

To build socialism in the rural south, as in the north, the post-1975 government called for socialist transformation to establish large-scale production in agriculture. This transformation had two main components: land redistribution and collectivisation. The former was considered a temporary measure while the latter was the end goal of socialist revolution in the countryside.

Post-1975 land reform in the south

Despite inheriting a land tenure system that was more or less equitable and dominated largely by middle peasants, the VCP decided to carry out land reform in the south. It gave several reasons for this reform. First, it was aimed at eliminating the social base of potential opponents such as landlords, rural capitalists and rich peasants. Resolution numbers 247/NQ-TW (29 September 1975) and 254/NQ-TW (15 July 1976) and Directive No. 235/CT-TW (20 September 1976) stressed the 'elimination of the vestiges of colonist and feudal exploitation of land', 'nationalisation of farms and the land of foreign capitalists', and the 'expropriation of farms, the lands of comprador capitalists, and treacherous landlords' and of landlords who had fled abroad.

Second, it was aimed at fulfilling the promise of the LTTT program, which the party used to attract peasants' support during the war. The party called this 'completing the remaining task of land revolution in the south' (*hoàn thành nhiệm vụ cách mạng ruộng đất ở Miền Nam*).[26]

Third, the party also wanted land reform to resolve postwar economic problems, especially food shortages. After Vietnam's reunification, both China and the United States cut their food aid to the country, so the party had to make food security and self-sufficiency top priorities. Land reform therefore aimed to increase food production and facilitate solidarity among peasants. The party's Directive No. 235 (dated 20 July 1976) stated:

---

26   ĐCSVN, Resolution No. 247/NQ-TW, pp. 368–406; ĐCSVN (2004), Nghị quyết của Bộ chính trị số 254/NQ-TW (ngày 15 tháng 7 năm 1976) về những công tác trước mắt ở Miền Nam [Politburo Resolution No. 254/NQ-TW (15 July 1976) on ongoing work in the south], in ĐCSVN, *Văn Kiện Đảng Toàn Tập: Tập 37, 1976* [*Party Document: Volume 37, 1976*], Hà Nội: NXB Chính Trị Quốc Gia, pp. 200–25; Ban Bí Thư (BBT) (1976), Chỉ thị 235/CT-TW *của Ban bí thư Trung ương Đảng Cộng Sản Việt Nam (ngày 20 tháng 9 năm 1976) về việc thực hiện nghị quyết của Bộ chính trị về vấn đề ruộng đất ở Miền Nam* [*Directive No. 235 of the Secretariat of the Central Committee Communist Party of Vietnam on the Implementation of the Politburo's Land Resolution in the South*], 20 September, Hà Nội: Ban bí thư Trung ương Đảng Cộng Sản Việt Nam, p. 2 (provided by a former staffer of the Central Committee for Agricultural Transformation in the south).

> Resolving the land problem in the south is aimed at not only eliminating the vestiges of feudal and colonist exploitation and making the landless and the land-short have means of production to make a living but also facilitating peasant solidarity and production … [Therefore,] in areas where the land problem is basically resolved, [we] can just carry out land reform in some necessary cases, not undo and do it again. In areas with vestiges of feudal and colonist exploitation, [we] will attempt to address that fast, definitely by 1976. Note that when sharing land to peasants, [we] must avoid dividing land into small parcels that are unfavourable for production.[27]

Finally, as in other socialist countries, Vietnam's post-1975 land reforms were temporary and preparatory measures towards collectivisation.

However, instead of taking a more radical approach, as in the north in 1953–57 (and as occurred in China in the 1950s), the government of reunified Vietnam took a more moderate and gradual approach in the south, although the content and emphasis of these reforms varied over time.[28] There are at least three main reasons the VCP chose this approach.

First, according to VCP accounts, party leaders recognised the positive legacy of previous agrarian reforms and admitted that 'the landlord class had already been largely eliminated' and 'the majority of land now belonged to peasants'.[29] One party scholar also noted that, by 1975–76, thanks to land reforms between the 1950s and the early 1970s, middle peasants (*trung nông*) made up the majority of farming households.[30] So, although party leaders knew that land and machinery were not distributed equally, especially in the Mekong Delta, the tenancy problem had already been largely eliminated in the south and certainly was not as serious as it had been in the north during the 1950s.[31] A government report said tenancy remained a problem in only a few

---

27 BBT, *Directive No. 235*.

28 Nolan, P. (1976), Collectivization in China: Some comparisons with the USSR, *The Journal of Peasant Studies* 3(2): 192–220, at p. 203. From 1975 to 1978, the authorities emphasised eliminating exploitation ahead of land redistribution. However, when collectivisation in the Mekong Delta failed to achieve its expected goal, the party attributed the failure to the incompleteness of land reform. Land redistribution (*điều chỉnh ruộng đất*) was therefore given prominence in the early 1980s, touching not only upper–middle peasants, but also middle peasants.

29 ĐCSVN, Resolution No. 247/NQ-TW, p. 382.

30 Lâm Quang Huyên (1985), *Cách Mạng Ruộng Đất ở Miền Nam Việt Nam* [*The Land Revolution in South Vietnam*], Hà Nội: NXB Khoa Học Xã Hội, p. 189.

31 Lê Duẩn (1980), *Cải Tạo Xã Hội Chủ Nghĩa ở Miền Nam* [*Socialist Transformation in the South*], Hà Nội: NXB Sự Thật, p. 74.

rural areas that had previously been under the prolonged control of Saigon's government; the tenanted land accounted for only about 1 per cent of total agricultural land, and the land rent was about 20 *gịa* (400 kg) of paddy per hectare.[32]

Second, as well as fulfilling its political objectives, the party tried to minimise any negative economic effects of reform. A radical reform program could cause chaos and a significant fall in food production. This is why VCP leaders often emphasised land reform be carried out by 'negotiating with each other', 'helping and unifying each other', 'being affectionate and rational' (*có tình có lý*) and 'allowing cultivators to continue to farm on parts of their current land' (*giữ nguyên canh là chính*). Party leaders argued that this approach could avoid disrupting agricultural production, strengthen peasants' solidarity and make collectivisation easier.[33]

Finally, although I found little evidence in official documents, it seems the party had learned a costly lesson from the radical land reforms in the north and did not want to repeat it in the south.[34] Party leaders still classified rural capitalists, rich peasants and upper–middle peasants as the 'exploiting' class and considered wage labour a capitalist form of exploitation.[35] Party leaders retained their objective to eliminate the enemies of socialism in the south, as they had in the north in the 1950s, but this was to be achieved in a quieter and more gradual way. The method was similar to that of re-education camps from which thousands of ex-government officials and supporters were quietly and discreetly sent to prison.[36]

---

32   Ban Cải Tạo Nông Nghiệp Miền Nam [hereinafter BCTNNMN] (1984), Báo cáo tình hình ruộng đất *và quá trình điều chỉnh ruộng đất trong nông thôn Nam Bộ* [*Report on Land Redistribution in the Southern Region*], January, Hồ Chí Minh: Ban Cải Tạo Nông Nghiệp Miền Nam, p. 3. *Gịa* is often used to measure paddy weights in the Southern Region (*Nam Bộ*). It is equal to 20 kg; therefore, 20 *gịa* is equal to 400 kg (ĐCSVN, Politburo Resolution No. 254/NQ-TW, pp. 214–16).
33   BBT (2004), *Directive No. 235*, pp. 279–80.
34   I found little evidence of this in official documents. In interviews, however, many local cadres, including former cadres of the Central Committee for Agricultural Transformation in the south, said these things to me (Fieldwork in Vietnam, May–December 2005).
35   ĐCSVN, Resolution No. 247/NQ-TW.
36   Thompson, L. C. (2010), *Refugee Workers in the Indochina Exodus, 1975-1982*, Jefferson, NC: McFarland & Co.

Socialist transformation of agriculture for collectivisation

Because the post-1975 agrarian reform was a key component of the socialist revolution, it is difficult to separate it from other economic and social reforms. To understand the rationale for or original objectives of this transformation in the south, it is necessary to examine them in the context of the overall socialist revolution.

According to party accounts—such as the second plenum's resolution of the fourth Communist Party Congress, the Secretariat's Directive No. 15/CT-TW (4 August 1977) and the Politburo's Directive No. 43/CT-TW (14 April 1978)—the objectives of socialist transformation in agriculture, or agricultural collectivisation, included 'taking agriculture into socialist large-scale production'; 'eliminating exploitation and the causes of exploitation, backwardness and poverty'; 'facilitating the collective mastery of the labouring people and developing agricultural production'; 'building up technical bases and bringing advanced science and techniques into production to increase productivity'; 'improving step by step the living standards of the peasants and constructing a new way of life in rural areas'; 'contributing to the reorganisation of productive forces at the national level'; and 'contributing to meeting the essential requirements for food and food stuff, industrial inputs and exports, and making agriculture a favourable base for socialist industrialisation'. The following sections will pinpoint in detail each of these objectives.[37]

---

37 ĐCSVN (2004), Nghị quyết Hội nghị lần thứ hai của Ban chấp hành Trung ương Đảng khóa IV, số 03/NQ-TW (ngày 19 tháng 8 năm 1977) [Resolution No. 03/NQ-TW of the Second Plenum of the Central Committee of the Party IV (19 August 1977)], in ĐCSVN, *Văn Kiện Đảng Toàn Tập: Tập 38, 1977* [*Party Document: Volume 38, 1977*], Hà Nội: NXB Chính Trị Quốc Gia, p. 298; ĐCSVN (2005), Chỉ thị của Ban bí thư số 15/CT-TW (ngày 4 tháng 8 năm 1977) về việc thí điểm cải tạo xã hội chủ nghĩa ở Miền Nam [Secretariat's Directive No. 15/CT-TW (4 August 1977) on experimenting with socialist agricultural transformation in the south], in ĐCSVN, *Văn Kiện Đảng Toàn Tập: Tập 38, 1977* [*Party Document: Volume 38, 1977*], Hà Nội: NXB Chính Trị Quốc Gia, pp. 741–7; ĐCSVN (2004), Chỉ thị của Bộ chính trị, số 43/CT-TW (ngày 14 tháng 4 năm 1978) về việc nắm vững và đẩy mạnh công tác cải tạo nông nghiệp ở Miền Nam [Politburo's Directive No. 43 on intensifying agricultural transformation in the south], in ĐCSVN, *Văn Kiện Đảng Toàn Tập: Tập 39, 1978* [*Party Document: Volume 39, 1978*], Hà Nội: NXB Chính Trị Quốc Gia, pp. 183–91.

From small and spontaneous to large-scale and planned production

The resolution of the fourth Communist Party Congress highlighted the role of agriculture in Vietnam's new era: 1) producing sufficient food for the consumption needs of the whole society and for food reserves; 2) supplying raw materials for industrialisation; and 3) producing for export.[38] The party believed collectivisation and collective ownership would make it easy to plan production at regional and national levels. It would also be easy to construct large areas of concentrated and specialised agricultural production. Through large-scale production, it would be possible for agriculture to adopt new and modern techniques and use science to push intensive farming (*thâm canh*), increase the number of crops per year (*tăng vụ*), expand cultivated areas (*mở rộng diện tích*), expand irrigation (*thuỷ lợi hoá*), increase mechanisation (*cơ giới hoá*) and adopt new seeds (*giống mới*).[39] The combination of all these factors could give Vietnam a modern and productive agricultural sector that guaranteed sufficient food production for the whole society plus surplus for industrialisation.

According to party leaders, the south would play an important role in fulfilling these new tasks because it possessed an abundance of fertile land, farm equipment and skilled labour, especially in the Southern Region (*Nam Bộ*).[40] The south, according to one study, had about 3.2 million hectares of cultivated land compared with 2 million hectares in the north. Moreover, the south had the potential to expand its agricultural land to 10 million hectares, compared with 4 million hectares in the north. Of this, the Mekong Delta would be able to extend agricultural production to 1,032,000 additional hectares of land; the South-Eastern Region to 779,000 hectares; the Central Highlands, 1,366,000 hectares; and Zone V of the Central Coast, 652,000 hectares.[41]

---

38 ĐCSVN, Resolution No. 03/NQ-TW of the Second Plenum.

39 ĐCSVN (2004), Nghị Quyết của Đại hội Đảng lần thứ IV của Đảng Cộng Sản Việt Nam (ngày 20 tháng 12 năm 1976) [Resolution of the Fourth Party Congress of the Communist Party of Vietnam (20 December 1976)], in ĐCSVN, *Văn Kiện Đảng Toàn Tập: Tập 37, 1976* [*Party Document: Volume 37, 1976*], Hà Nội: NXB Chính Trị Quốc Gia, pp. 930–2.

40 Phan Văn Đáng (1978), Tập dượt đi lên hợp tác xã nông nghiệp [Experiment with agricultural collectives], in Võ Chí Công et al. (eds), *Con đường làm ăn tập thể của nông dân* [*The Collective Farmer's Way*], Hồ Chí Minh: NXB Tổng Hợp Thành Phố, (TP) Hồ Chí Minh, p. 110.

41 Nguyễn Trần Trọng (1980), *Những vấn đề công tác cải tạo và xây dựng nông nghiệp ở các tỉnh phía Nam* [*Ongoing Tasks for Transforming and Building the South's Agriculture*], Hà Nội: NXB Nông Nghiệp, p. 182.

Despite placing a high value on the south's agricultural potential, party leaders strongly criticised it for 'individual farming', 'fragmented landholding', 'unequal development' and the influence of capitalism.[42] They argued 'the fragmentation of agricultural production results from small-scale production, individualised farming aimed at fulfilling subsistence and narrow demands of local markets'.[43] Individual farming was 'spontaneous, unplanned' (*tự phát, tuỳ tiện*) and 'fragmented' (*manh mún*). It had 'a low level of specialisation and cooperation' and was 'technically backward'.[44] Moreover, the individual farming system, party leaders believed, had 'backward' production relations that hindered adoption of modern techniques and better use of land.[45]

In general, according to the party, the south had high agricultural potential that had not been fully exploited. The task was therefore to transform the old system of agriculture into a new one of 'planned, concentrated and large-scale production', 'specialisation' (*chuyên môn hoá*), 'cooperativisation' (*hợp tác hoá*), 'linkagisation' (*liên hiệp hoá*) and collectivisation.[46]

Eliminating exploitation and its causes, poverty and backwardness

In the view of party leaders, land redistribution would provisionally eliminate exploitation in farming but would not eliminate the causes of exploitation.[47] A party scientist even argued that 'eliminating the feudal land tenure system and implementing the slogan "land to the tillers"

---

42   Ibid.

43   Hồng Giao (1984), *Đưa Nông nghiệp lên một bước lớn Xã hội chủ nghĩa* [*Taking Agriculture One Step Towards Socialist Large-Scale Production*], Hà Nội: NXB Sự Thật, p. 23.

44   Trần Văn Doãn (1986), *Như thế nào là nông nghiệp một bước lên sản xuất lớn xã hội chủ nghĩa* [*What is One Step of Agriculture Towards Socialist Large-Scale Production*], Hà Nội: NXB Nông Nghiệp, p. 5.

45   Võ Văn Kiệt (1985), *Thực hiện đồng bộ ba cuộc cách mạng ở nông thôn* [*Simultaneous Execution of Three Revolutions in Rural Areas*], Hồ Chí Minh: NXB Tổng Hợp TP Hồ Chí Minh, pp. 47, 128; Nguyễn Trần Trọng, *Ongoing Tasks*, p. 9.

46   Nguyễn Trần Trọng, *Ongoing Tasks*, p. 9; Tố Hữu (1979), Phát động phong trào quần chúng thực hiện thắng lợi công cuộc cải tạo xã hội chủ nghĩa đối với nông nghiệp Miền Nam [Campaign to succeed in socialist agricultural transformation in the south], in Võ Chí Công et al. (eds), *Khẩn trương và tích cực đẩy mạnh phong trào hợp tác hóa nông nghiệp Miền Nam* [*Urgently and Positively Promote the Acceleration of Collectivisation in the South*], Hà Nội: NXB Sự Thật, p. 42.

47   Kerkvliet, *The Power of Everyday Politics*, p. 10. The party believed without collectivisation a few successful farming households would end up owning much of the land, undermining the ideal of social and economic equality.

were in fact beneficial to the development of capitalism in rural areas.'[48] Rural households would soon become unequal. Replacing private ownership with collective ownership would guarantee the elimination of exploitation and its causes as well as poverty and backwardness.[49] At the second plenum of the fourth party congress (6–16 December 1977), in assessing the achievements and the shortcomings of the past 20 years of collectivisation in the north, Lê Duẩn minimised the failure of the north to increase productivity and the living standards of peasants. Instead, he emphasised the achievements. Collectivisation in the north, he said, had eliminated the cause of class conflicts in rural areas, facilitated solidarity among different rural groups (such as religious and non-religious people, Kinh people and ethnic minorities) and protected the livelihoods of people, especially the elderly, infants, invalids and the families of war martyrs (*gia đình thương binh liệt sĩ*). Second, it had improved irrigation, facilitated new farming techniques and increased the number of crops harvested per year. All these factors led to 'an increase in productivity and food production in the north despite still facing stressful food shortage in the time of calamity'. Third, it changed the face of rural society; thanks to collectivisation, cultural, education, healthcare and material conditions in rural areas had gradually improved. Finally, it played an essential role in contributing to the defeat of the American invasion and saving the country. He believed collectivisation in the south could achieve similar results.[50]

Backing socialist industrialisation and ensuring food security

The leaders of some socialist countries, such as Russia and China, considered agriculture to be a source of financial surpluses for carrying out industrialisation and collectivisation as the keys to state-centred accumulation and the primacy of the growth of heavy industry.[51] Post-1975 agrarian reform in Vietnam was also aimed at supporting industrialisation as well as ensuring food security for the whole society.

---

48 Nguyễn Huy (1985), *Mấy vấn đề lý luận và thực tiễn của cách mạng quan hệ trong nông nghiệp nước ta* [*Theories and Practices of Revolution in the Production Relations of Our Country's Agriculture*], Hà Nội: NXB Khoa Học Xã Hội, p. 121.

49 Đại Đoàn Kết (1977), Nghị quyết lần thứ II: Ban chấp hành Trung ương Đảng khóa IV ra nghị quyết [Resolution II of the Central Committee of the Party IV], *Đại Đoàn Kết*, 3 September 1977, p. 11.

50 ĐCSVN (2004), Đề Cương kết luận của đồng chí Lê Duẩn tại Hội nghị lần thứ II [Lê Duẩn's final statements at second plenum], in ĐCSVN, *Văn Kiện Đảng Toàn Tập: Tập 38, 1977* [*Party Document: Volume 38, 1977*], Hà Nội: NXB Chính Trị Quốc Gia, pp. 254–5.

51 Selden, M. (1994), Pathways from collectivization: Socialist and post-socialist agrarian alternatives in Russia and China, *Review (Fernand Braudel Center)* 17(4), pp. 423–49, at p. 425; ĐCSVN, Resolution of the Fourth Party Congress, p. 917.

During the war, both the north and the south had relied heavily on foreign aid, including food.[52] However, soon after the war, this aid was gradually cut off or significantly reduced, and some imported foods were no longer available.[53] Ensuring food for the whole society therefore became a top concern of the VCP. With collectivisation, party leaders believed, Vietnam could deal with its food shortage. Moreover, collectivisation would create conditions in which 'every labourer has a job, every field is properly used and every industry can develop'.[54]

Controlling rural society and consolidating power

Party leaders attached great importance to controlling rural areas in times of war and post war. In wartime, within a competitive environment and focused on winning the war, the party had adopted policies favouring peasants' interests, which Brantly Womack calls 'mass-regarding in policy'.[55] However, reunification changed the context in terms of not only power relations between the party and the peasants, but also the main concerns of the party. Although the party still paid attention to peasants' interests, its ideology favoured other matters, too—such as controlling society, land, labour, production and grains to strengthen its socialist building projects.

In other words, the primary concern of the leaders of reunified Vietnam was to control the south politically, economically and socially, to consolidate their power and reorganise production according to their socialist blueprint. Party leaders often called for a strategy of 'holding firmly to the proletariat dictatorship' to control and manage all aspects of society and the economy, monitoring people's political, economic, cultural and social activities.[56] Socialist transformation included eliminating the political, social and economic bases of all perceived opposition classes. Revolutionary leader Võ Văn Kiệt pointed out:

52 Vo Nhan Tri (1990), *Vietnam's Economic Policy Since 1975*, Singapore: Institute of Southeast Asian Studies, p. 58.
53 ĐCSVN (2005), Chỉ thị của Ban bí thư số 02/CT-TW (ngày 21 tháng 1 năm 1977) về những việc trước mắt để giải quyết lương thực [Directive of the Secretariat No. 02/CT-TW (21 January 1977) on immediate matters for food processing], in ĐCSVN, *Văn Kiện Đảng Toàn Tập: Tập 38, 1977* [*Party Document: Volume 38, 1977*], Hà Nội: NXB Chính Trị Quốc Gia], p. 2.
54 Hồng Giao, *Taking Agriculture One Step*, pp. 26–7.
55 Womack, B. (1987), The party and the people: Revolutionary and postrevolutionary politics in China and Vietnam, *World Politics* 39(4), pp. 479–507.
56 ĐCSVN (2004), Báo cáo tổng kết công tác xây dựng Đảng, và sửa đổi điều lệ Đảng (ngày 17 tháng 12 năm 1976) [Report on building party organisation and changing party regulations (17 December 1976)], in ĐCSVN, *Văn Kiện Đảng Toàn Tập: Tập 37, 1976* [*Party Document: Volume 37, 1976*], Hà Nội: NXB Chính Trị Quốc Gia, pp. 749, 789.

[T]hrough economic transformation the state consolidates and strengthens the proletariat dictatorship and collective mastery of labouring people, roots out completely counter-revolutionary forces, completes the economic unification of the country and facilitates the entire strength of the socialist state.[57]

One of the objectives of socialist agricultural transformation was to bind peasants with the party-state to isolate perceived opposition groups and gain social control of the countryside. Moreover, party leaders believed that controlling peasants, their production and their produce would help them also control non-staple food producing groups and their goods in the cities. Lê Duẩn argued:

[I]f the state controls staple food, it can control industrial goods … controlling staple food means controlling everyone's essential goods which enables control of the products of large industries, small industries and handicraft producers.[58]

During my interviews, some of my respondents told me the communists controlled people by controlling their stomachs. Therefore, controlling the countryside and food production became important to the party in the post-reunification period. At the fourth party congress in December 1976, Premier Phạm Văn Đồng stressed:

[I]n agriculture, be quick to cut off the relationship between the capitalists and the peasants, organise immediately the relationship between the state and the peasants, using this relationship to help peasants develop production and request them to sell food to the state.[59]

Another objective of socialist agricultural transformation was to select and purify local cadres to consolidate the power of the party-state in the rural south. During the war, many southern party cells had been destroyed. Others, especially in the 'religious areas' of the Mekong Delta, barely functioned and were considered 'thin' (*cơ sở đảng mỏng*) or 'blank' (*cơ sở đảng trắng*). Thus, party leaders called for a consolidation of the party's base and recruitment of new members in the south,

---

57 Võ Văn Kiệt, *Simultaneous Execution of Three Revolutions*, p. 40.
58 Lê Duẩn, *Socialist Transformation in the South*, p. 14.
59 ĐCSVN (2004), Phương hướng nhiệm vụ và mục tiêu chủ yếu của kế hoạch 5 năm 1976–1980 [Key tasks and objectives of the five-year plan, 1976–1980], in ĐCSVN, *Văn Kiện Đảng Toàn Tập: Tập 37, 1976* [*Party Document: Volume 37, 1976*], Hà Nội: NXB Chính Trị Quốc Gia, p. 655.

as well as the building and consolidation of political organisations, the testing of cadres and purification of 'bad elements' within state and party organisations.[60]

The content and steps of socialist transformation of agriculture

According to VCP leaders, the purpose of the socialist transformation of agriculture was to carry out three intertwined revolutions in the countryside: a revolution in production relations, a revolution in science and technology and a revolution in thought and culture. At the fourth congress in December 1976, Lê Duẩn emphasised that the socialist transformation was to 'combine a revolution in production relations with a revolution in technology and science and in thoughts and culture, as well as reorganising national production and circulation'.[61]

The overall aim of the 'three revolutions' in the rural south was to create 'a regime of socialist collective ownership, socialist large-scale production, adoption of new technology and socialist men with socialist values and culture'.[62]

The following sections describe these three revolutions and the steps towards socialist agricultural production.

Revolution in production relations

Post-1975 land reform was considered part of the process of establishing collective farming. Therefore, its beneficiaries would have the right to use but not to own the land. 'The state', according to one party resolution, 'does not provide a certificate of land ownership to the beneficiaries'.[63]

---

60 ĐCSVN (2004), Chỉ thị của Ban bí thư, số 273/CT-TW (ngày 24 tháng 9 năm 1976) về việc củng cố tổ chức cờ sở Đảng và kết nạp Đảng viên mới ở Miền Nam [Secretariat's Directive No. 273 on consolidating party organisation in the south], in ĐCSVN, *Văn Kiện Đảng Toàn Tập: Tập 37, 1976 [Party Document: Volume 37, 1976]*, Hà Nội: NXB Chính Trị Quốc Gia, p. 285.
61 Đại Đoàn Kết (1977), Nhiệm vụ cải tạo quan hệ sản xuất Miền Nam [Ongoing task for socialist transformation in the south], *Đại Đoàn Kết*, 17 September 1977, p. 14.
62 ĐCSVN (2004), Báo cáo chính trị của Ban chấp hành Trung ương Đảng tại Đại hội đại biểu toàn quốc lần thứ IV, do đồng chí Lê Duẩn trình bày [Political report of the Party Executive Committee at the fourth national representative meeting], in ĐCSVN, *Văn Kiện Đảng Toàn Tập: Tập 37, 1976 [Party Document: Volume 37, 1976]*, Hà Nội: NXB Chính Trị Quốc Gia, pp. 454–608.
63 Hội Đồng Chính Phủ [hereinafter HĐCP] (1976), *Quyết định số 188/CP của Hội Đồng Chính Phủ (ngày 25 tháng 9 năm 1976) về chính sách xóa bỏ tàn tích chiếm hữu ruộng đất và các hình thức bóc lột thực dân, phong kiến ở Miền Nam Việt Nam* [*Ministerial Council's Decision No. 188/CP (25 September 1976) on the Policy of Eliminating Land Tenure and Other Forms of Colonial and Feudal Exploitation in the South*], Hà Nội: Hội Đồng Chính Phủ, p. 7.

Party leaders envisioned collectivisation requiring a prolonged political campaign and class struggle between capitalism and socialism.[64] To ensure its success, the party called for a 'step-by-step' approach to move from low to high and from simple to complicated forms of collective farming that were suitable for each region.[65] The change meant moving from 'simple interim forms of collective organisation'—production solidarity teams (*tổ đoàn kết sản xuất*) or labour exchange teams (*tổ đổi công vần công*)—to a medium form of collective (production units: *tập đoàn sản xuất*) and then to full collectives (*hợp tác xã*). This process is quite similar to that of collectivisation in the north, which shifted from 'mutual aid teams' (*tổ đổi công*) to low-level collectives (*hợp tác xã bậc thấp*) and then to high-level collectives (*hợp tác xã bậc cao*).[66]

The resolution of the fourth party congress in 1976 was to set 1980 as the target date for the completion of agricultural transformation in the south, bringing most peasant households and their land into the collectives. In addition, VCP leaders planned to establish state farms that would occupy about one-third of cultivated areas and become dominant in production and distribution. Party leaders envisioned state farms (*nông trường quốc doanh*)—so-called agro-industrial corporations (*tổ hợp công nông nghiệp*)—being the largest production organisation in the socialist agricultural sector. They would operate like an industrial factory, relying on mechanisation, specialisation and the use of intensive farming techniques. The state farms, according to party leaders, would be expected to set a good example for agricultural collectives in the use of scientific methods of management and farming. Individual farming would be eliminated or become insignificant.

---

64 Phạm Văn Kiết (1978), Nông dân đang sôi nổi đi lên làm ăn tập thể [Peasants are eager for collective farming], in Võ Chí Công et al. (eds), *Con đường làm ăn tập thể của nông dân* [*The Collective Farmer's Way*], Hồ Chí Minh: NXB TP Hồ Chí Minh, p. 20.

65 ĐCSVN, Politburo's Directive No. 43, p. 188.

66 Quang Truong, Agricultural collectivization, p. 56; Nguyễn Huy, *Theories and Practices of Revolution*, pp. 95–6. Another interim form was the farming machine team (*tổ hợp máy nông nghiệp*) established in the Mekong Delta and South-Eastern Region. Each team had five to seven peasants who possessed farm machines. Depending on the classification of the machines (large, medium or small), these teams were organised under the direct leadership of the agricultural department of the district, the commune or the hamlet's production department (*Ban sản xuất ấp xã*), respectively. These organisations were supposed to be 'interim' or 'transitional' (*qúa độ*) steps in establishing collective machine units (*tập đoàn máy*) under the control of the district's authorities or specialised machine teams (*đội chuyên máy*) under the control of collectives or production units.

Revolution in science and technology

While the revolution in production relations was to see the creation of socialist large-scale production organisations (collectives and state farms), the revolution in science and technology would modernise agriculture, which was considered essential to making collective farming superior to individual farming. Party leaders therefore stressed, as well as carrying out collectivisation in the south, 'the need to combine collectivisation with extending irrigation [*thuỷ lợi hoá*] and mechanisation [*cơ giới hoá*] and using modern and advanced science and techniques for cultivation and animal husbandry'.[67]

Irrigation

In assessing the irrigation systems of the south, the party concluded they were too few, too small and often individually owned. War had also destroyed some. Party leaders planned to double the amount of irrigation by 1980.[68]

In the Mekong Delta, the party's irrigation program emphasised improving the canal systems by dredging existing channels and making new ones, and building new irrigation systems to treat acid sulphate soil (*rửa phèn*) and retarding floods and salinity intrusion (*chống lũ, chống mặn*).[69] For the Central Coast, which had only 460,000 hectares of agricultural land in mid-1970 and where farmers relied heavily on rainfall because irrigation systems were poor, the party called for the repair of existing canals and the establishment of more dykes and water-pumping stations.[70] By 1980, the Central Coast was expected to irrigate 180,000 to 200,000 hectares of double-cropped rice fields.[71]

---

67   ĐCSVN, Key tasks and objectives of the five-year plan, p. 626.
68   ĐCSVN, Resolution No. 03/NQ-TW of the Second Plenum, p. 315.
69   There were four provinces in the Central Coast Region: Quảng Nam-Đà Nẵng, Nghĩa Bình, Phú Khánh and Thuận Hải (Nguyễn Dương Đáng. (1983). *Kinh tế nông nghiệp Xã hội chủ nghĩa* [*Economics of Socialist Agriculture*]. Hà Nội: NXB Nông Nghiệp, p. 105).
70   ĐCSVN (2004), Báo cáo của Bộ chính trị tại hội nghị lần thứ hai Ban chấp hành Trung ương khóa IV [Report of the Politburo at the Second Conference of the Central Committee, Session IV], in ĐCSVN, *Văn Kiện Đảng Toàn Tập: Tập 38, 1977* [*Party Document: Volume 38, 1977*], Hà Nội: NXB Chính Trị Quốc Gia, p. 180.
71   Nguyễn Dương Đáng, *Economics of Socialist Agriculture*, p. 42; Nguyễn Trần Trọng, *Ongoing Tasks*, p. 328.

Mechanisation

For party leaders, mechanisation meant substituting machinery for animal and human power so as to increase productivity and efficiency.[72] The authorities also believed mechanisation would help to attract peasants, especially middle ones, to collective farming because many farming households in the Southern Region were already using some machinery.[73] One party scientist argued that, without 'combining collectivisation with mechanisation, attracting peasants into collectives will be difficult' because it would not be able to demonstrate 'its superiority over individual farming'.[74]

To utilise existing agricultural machinery in the south, party leaders urged each district to organise privately owned machines into machinery teams (*tổ hợp máy*), machinery units (*tập đoàn máy*), specialised machinery teams (*đội máy chuyên doanh*) and collective machinery teams (*đội máy tập thể*). Each district was also supposed to build state machinery stations (*trạm máy quốc doanh*) equipped with 'large' machines (*máy lớn*) supplied by the state or bought from individuals.[75] At the second plenum of the fourth party congress, VCP leaders planned to import 18,700 large tractors, 30,000 small ones and other machinery to increase the mechanisation rate in land preparation to 50 per cent for the whole country and 74 per cent for the Mekong Delta.[76]

Chemical inputs and new seeds

Before reunification, southern peasants, especially in the Mekong Delta, had used chemical fertilisers. The importation of such fertilisers in the south had increased dramatically after 1960 and reached 372,183 tonnes in 1973. The average amount of chemical fertiliser used per hectare of agricultural land reached about 120 kilograms. The greater use of fertilisers was associated with an increased adoption of new rice seeds (*lúa thần nông*) in the south, which were planted on 41,000 hectares in 1968 (accounting for 1.4 per cent of rice-growing land) and on 890,400 hectares in 1973 (31 per cent). However, the adoption of new seeds in the south was low compared with a rate of 60 per cent in the north at the same time—which party leaders used to indicate the

---

72   ĐCSVN, Resolution No. 03/NQ-TW of the Second Plenum, p. 232.
73   Nguyễn Trần Trọng, *Ongoing Tasks*, p. 246.
74   Nguyễn Huy, *Theories and Practices of Revolution*, p. 96.
75   ĐCSVN, Resolution No. 03/NQ-TW of the Second Plenum, p. 210.
76   Ibid., p. 178.

superiority of socialist agriculture. As well as the low rate of new seed adoption, party leaders criticised southern farming for using too little fertiliser, especially compared with the north of the country. They also urged rural southerners (families, collectives and state farms) to make 'green manure' (*làm phân xanh*) and 'dung manure' (*phân chuồng*).[77]

The party believed that, by implementing irrigation and mechanisation and the adoption of new seeds and fertilisers, the state could gradually control peasants' production and shift them into collective organisations.

Revolution in thought and culture

Taking the south into socialism, carrying out socialist industrialisation and establishing large-scale socialist production of agriculture were decisions made by the top party leaders. To see their policies carried out, they needed people's participation, conformity and endorsement. The party realised that southerners had long engaged in capitalist production, had a private ownership mindset (*đầu óc tư hữu*) and had capitalist tendencies (*có khuynh hướng tư bản chủ nghĩa*). Leaders also believed the harmful legacies of two decades of US neo-colonialism posed great obstacles to the construction of socialism in the south.[78] For example, they believed colonialism and bourgeois thoughts (*tư tưởng tư sản*) were entrenched in the south and 'anti-revolutionary' groups (*bọn phản cách mạng*) were still active.

Therefore, the socialist revolution to determine 'who would triumph over whom', to transform private into public ownership and to replace individual with socialist large-scale production would encounter strong resistance.

To tackle this situation, party leaders set out to transform people's thoughts and culture to fit their policies. They called this effort the 'revolution in thought and culture'. A prevalent guiding slogan was a statement by President Ho Chi Minh: 'The first and essential condition for constructing socialism is to have socialist people' (*muốn xây dựng chủ nghĩa xã hội, trước hết cần có những con người xã hội chủ nghĩa*).[79]

---

77 Taylor, P. (2001), *Fragments of the Present: Searching for Modernity in Vietnam's South*, Honolulu: University of Hawai'i Press, p. 31.

78 Hồ Chí Minh's statement on new socialists was cited in Lê Duẩn's report at the first meeting of the unified National Assembly, on 25 June 1976 (ĐCSVN, Political report of the Party Executive Committee, p. 151).

79 Ibid.

Socialist people were supposed to have the following characteristics: 1) correct thoughts and affection, adequate knowledge and the ability to undertake collective mastery over society, the natural world and oneself (*làm chủ xã hội, thiên nhiên và bản thân*); 2) high levels of volunteerism and a determination to overcome every difficulty to complete assigned tasks; 3) be honest, disciplined, skilful and productive, love working and detest living off others (*ăn bám*), and have respect for and protect public property; and 4) love socialism and have the pure spirit of the 'international proletariat' (*quốc tế vô sản*).[80]

To produce these kinds of 'new socialists', party leaders called for multiple measures involving education, administration, political and cultural activities, coercion and economic incentives.[81] Socialist people were created not only in the Communist Party, but also in 'every mass, economic, cultural and social organisation, in every industry and at every level of administration, in every town and village and family'.[82] Central government and local newspapers, socialist literature and the arts were also required to serve the construction of new socialist people by 'praising good people and good merits' (*ca ngợi người tốt việc tốt*) and criticising 'negative phenomena' (*hiện tượng tiêu cực*) in society and 'the legacy of feudalism and colonialism'.[83]

Party leaders considered local-level cadres the most important agents for the success of socialist transformation. They reasoned that agricultural collectivisation would transform the nature of production organisations and the way of life in the countryside. The requirements for managing socialist large-scale production were completely different to those of the small individual economy; therefore, cadres (including political, managerial and technical cadres) would be determining factors for success.[84]

In addition to the general characteristics of new socialists, cadres were supposed to be 'frugal' (*cần kiệm*), 'moral' (*liêm chính*), 'live simple, clean and sound lives', 'fight against privilege, embezzlement, collusion and trespassing on state property' and 'repel the influence

---

80   Ibid.
81   Ibid., pp. 500–1.
82   ĐCSVN, Resolution of the Fourth Party Congress, p. 935.
83   Nguyễn Trần Trọng, *Ongoing Tasks*, p. 277.
84   ĐCSVN, Report on building party organisation, pp. 743, 849.

of the bourgeois lifestyle'.[85] Through mass campaigns of agricultural transformation and collectivisation, local party cells (*đảng bộ cơ sở*) would be able to identify the 'good' and the 'bad' cadres. The former would include those who were committed to large-scale production (*quyết tâm theo con đường sản xuất lớn*). The latter would be those who still harboured the 'thoughts of peasants' (*tư tưởng nông dân*) and 'thoughts of self-satisfaction and longing for individual farming' (*luyến tiếc làm ăn riêng lẻ*).[86]

Party leaders also realised that, in addition to the influence of local cadres, peasants' attitudes, motivations and actions would significantly affect the results of socialist transformation in general and the performance of collective farming in particular. The results would be excellent if people 'absolutely trusted' the party's policies. Therefore, soon after reunification, the party tried to attract peasants in the south to join 'peasant associations' (*nông hội*) to educate them to 'enhance a patriotic spirit' (*nâng cao tinh thần yêu nước*) and 'love socialism' (*yêu chủ nghĩa xã hội*).[87] It would be important to instil in peasants 'socialist thought', to educate them about the party-state's policies, to give them a 'consciousness of building socialism' (*có ý thức xây dựng chủ nghĩa xã hội*) and to encourage them to perform their obligations to the state well (*thực hiện tốt nghĩa vụ với nhà nước*).[88]

Conclusion

Soon after reunification, the government in Hanoi decided to carry out a socialist revolution in the south to reunify the country politically, socially and economically. VCP leaders considered socialist agrarian reform a key component of the socialist revolution.

---

85 Ibid. A former district cadre in Chợ Mới said he learnt a lesson about cadres during the campaign of collectivisation: good cadres were those who 'took care of the people but did not follow the ideas of the masses' (*lo cho dân nhưng không chạy theo quần chúng*) (Author's interview, 17 June 2005, Chợ Mới).

86 ĐCSVN, Lê Duẩn's final statements at second plenum, p. 289.

87 ĐCSVN, Political report of the Party Executive Committee, p. 564.

88 ĐCSVN, Resolution No. 03/NQ-TW of the Second Plenum, p. 290; Phan Văn Đáng, Experiment with agricultural collectives, pp. 30–1.

Driven by Marxist–Leninist doctrine and high expectations of their capacity and the south's economic potential, the VCP leaders believed they could succeed in building a centrally planned economy, socialist industrialisation and large-scale production even though this had not been fully accomplished in the north. In the agricultural sector, this vision included two main components: land redistribution and collectivisation.

To ensure the success of collectivisation, VCP leaders paid great attention to its preparatory steps: redistributing land, bringing peasants into interim collective organisations, training cadres and building the capability of the local authorities. They called for the simultaneous execution of three revolutions—in production relations, in science and technology and in culture.

3

Postwar restoration and preparations for collectivisation

Introduction

Vietnam is one of the most bombed countries in world history. After three decades of war (1945–75), Vietnam inherited a devastated economy, society and ecology. Rural destruction in the southern half of Vietnam was especially severe, and thousands of villages were heavily affected by war. Millions of hectares of agricultural land were bombed repeatedly and, by 1975, according to a Vietnamese Communist Party (VCP) report, 560,000 hectares of cultivated land had been left untended.[1] One and a half million buffaloes and oxen were killed.[2]

The south faced another postwar problem: massive urban unemployment. During the conflict, large numbers of rural refugees were moved or fled to cities and towns, where they often worked in military-related sectors of the economy. At war's end, a majority of these refugees and soldiers and civil officials discharged by the government in Saigon were unemployed. According to VCP reports, the total urban population in 1975 was 7 million, of whom about 3 million (30 per cent) were unemployed.[3] So, after the war, central

---

1   ĐCSVN (2004), Báo cáo của Bộ chính trị tại Hội nghị Trung Ương Đảng lần thứ 24 [Report of the Politburo at the 24th Party Central Committee Conference], in ĐCSVN, *Văn Kiện Đảng Toàn Tập: Tập 36, 1975* [*Party Document: Volume 36, 1975*], Hà Nội: NXB Chính Trị Quốc Gia, p. 318.
2   Quang Truong, Agricultural collectivization, p. 155.
3   Ibid.

and local authorities emphasised the consolidation of political power and economic restoration. Inherent in these policies, however, was preparation for collectivisation. In other words, after the war, the VCP focused simultaneously on establishing its authority, restoring production, implementing land reform and solving other postwar problems.[4]

This chapter examines the implementation of these policies in the first few years after the war ended and prior to intense collectivisation of farming. In particular, the chapter focuses on the consequences of war and how local governments in parts of the Central Coast province of Quảng Nam-Đà Nẵng (QN-ĐN) and the Mekong Delta province of An Giang struggled to consolidate their political power and implement these policies and how local officials (who were policy implementers) and peasants reacted to them.

By comparing these two regions, the chapter reveals differences and similarities in their postwar conditions, which led to differences in the form and magnitude of peasants' and local cadres' politics and attitude to state policies. It shows variations in policy implementation and explains how local conditions affected the implementation and performance of national policies.

Postwar policies in the Central Coast

Rebuilding the war-torn economy

The Central Coast was the worst affected region in southern Vietnam in terms of lives lost and social, economic and ecological destruction. One province in that region was QN-ĐN (which was divided into two separate provinces in 1996). According to Quảng Nam's Department of Statistics, more than two-thirds of QN-ĐN's agricultural land was abandoned and uncultivated in 1975, thousands of people had been killed or injured and unexploded mines littered the countryside. More than three-quarters of all villages had been destroyed, forcing peasants to flee and live together in a few refugee areas, bringing economic activity to a standstill. Therefore, after the war, the province 'faced a severe food

---

4   See ĐCSVN, Resolution No. 247/NQ-TW.

shortage and acute unemployment'.[5] A large proportion of arable land was uncultivated. An article in the *Quảng Nam-Đà Nẵng* newspaper in December 1975 summarised the situation in verse: 'fields in rural areas lack draught animals; gardens were abandoned, houses were empty, and the people were prostrate and hungry' (*đồng quê vắng bóng trâu cày, vườn hoang nhà vắng dân gầy xác xơ*).[6]

Despite heavy destruction, the authorities in QN-ĐN rapidly consolidated their power in all parts of the province. By September 1976, according to the former provincial chairman of QN-ĐN:

> [A] complete system of revolutionary authority was quickly built from province to district, commune, ward, subcommune and subward. The revolutionary authorities swiftly controlled and managed all urban areas and large rural areas.[7]

At least three factors helped the new authorities in QN-ĐN consolidate their power. First, large rural parts of the province and the wider Central Coast region were under the influence of the Việt Minh during the war with France (1945–54) and then under the National Liberation Front (NLF) during the war against the US-backed government in Saigon (1954–75). Despite the 'liberated areas' (*vùng giải phóng*) being reduced significantly in the late 1960s, the NLF still controlled many remote and mountainous rural areas economically and politically. In the liberated areas, the NLF was able to carry out its policies and campaigns.[8] This familiarity and strong relationship with farmers enabled the new authorities to control and successfully deal with postwar society.

Second, QN-ĐN and other parts of the Central Coast region did not face huge problems filling government and party positions thanks to the large number of local revolutionaries who survived the war and others who returned there from northern Vietnam. During the war,

---

5   Cục Thống Kê tỉnh Quảng Nam [hereinafter CTKQN] (2005), *Quảng Nam 30 Năm Xây Dựng và Phát triển* [*Quảng Nam's Socioeconomic Development over the Past 30 Years*], Tam Kỳ: Cục Thống Kê tỉnh Quảng Nam, p. 22.

6   Đồng quê vắng bóng trâu cày, vườn hoang nhà trống dân gầy xác xơ [Fields in rural areas lack draught animals; gardens were abandoned, houses were empty, and the people were prostrate and hungry], *Quảng Nam-Đà Nẵng*, 15 December 1975, p. 1.

7   Hoàn thành thắng lợi vẻ vang nhiệm vụ xây dựng đất nước, xây dựng chế độ mới, con người mới Xã hội chủ nghĩa [Completing the task of building the country, the new regime and new socialist men], *Quảng Nam-Đà Nẵng*, 8 September 1976, p. 1.

8   Quyết thắng trên mặt trận nông nghiệp [Be determined to win on the agricultural front], *Quảng Đà*, 30 April 1974, p. 1; Lời kêu gọi ra sức gia tăng sản xuất, thực hành tiết kiệm [Do the best to increase food production and be thrifty], *Quảng Đà*, 30 April 1974, p. 2.

the NLF in QN-ĐN had recruited a large number of revolutionaries who operated locally or were sent to the north for training. Despite the surrender or killing of many revolutionaries from the late 1960s to the early 1970s, their numbers were still considerable. Quảng Nam's records show that, during the war, Bình Lãnh (a mountainous commune) suffered severe destruction. However, at least 25 revolutionary soldiers and 20 other revolution-supporting families still lived in the Bình Lãnh commune.[9] Likewise, Thăng Phước commune in Thăng Bình district was reportedly 'wiped clean' (bị xoá trắng) of its revolutionary base, but after the war, the number of surviving revolutionaries was sufficient to fill key positions in the communal and subcommunal authorities (chính quyền thôn).[10]

Finally, the flattened, war-torn society and economy made it somewhat easy for the new authorities to exert their power without confronting strong resistance from opposition groups.

Along with consolidating their power bases, the new authorities in QN-ĐN focused on solving the problems of refugees, unemployment and production. According to the Quảng Nam-Đà Nẵng newspaper, after the war, the province sent 400,000 refugees back to their home provinces (Quảng Trị and Thừa Thiên Huế). QN-ĐN also sent 700,000 refugees in urban areas back to their rural homes. In dealing with unemployment, the new authorities decided to move large numbers of unemployed people in urban areas either to the new economic zones in the Central Highlands or to rural areas.[11]

In rural areas, the new authorities focused on restoring agricultural production and preparing for collectivisation and socialist large-scale agriculture. This work included restoration of abandoned land and reclamation of new land (khai hoang, phục hoá), redistribution of landholdings (điều chỉnh ruộng đất), improvement of irrigation (làm thuỷ lợi), extension of cultivated areas (mở rộng diện tích canh tác), field transformation (cải tạo đồng ruộng) and intensive farming.

---

9   Tỉnh Ủy Quảng Nam (2003), Quảng Nam Anh Hùng, thời đại Hồ Chí Minh, Kỷ Yếu 6/2003 [Quảng Nam is a Hero in the Age of Ho Chi Minh], Tam Kỳ: Tỉnh Ủy Quảng Nam, pp. 319–21.
10   Toàn xã Thăng Phước làm ăn trong các tổ đổi công thường xuyên [The whole population of Thăng Phước commune is organised into regular labour exchange teams], Quảng Nam-Đà Nẵng, 23 May 1977, p. 4.
11   Nhân dân tỉnh ta chẳng những đánh giặc giỏi mà còn giàu nghị lực và tài năng sáng tạo trong xây dựng lại quê hương giàu đẹp [Our province's people fought the enemy and are building the country well], Quảng Nam-Đà Nẵng, 29 March 1976, p. 3.

Land restoration and redistribution

Land restoration

Soon after reunification, the new authorities in QN-ĐN launched a campaign to 'attack weeds in fields' (*chiến dịch tấn công đồng cỏ*) and 'remove unexploded landmines' (*tháo gỡ bom mìn*). The *Quảng Nam-Đà Nẵng* newspaper in March 1976 reported that authorities had mobilised thousands of urban youths to work in rural areas. In addition, they mobilised thousands of army engineers (*công binh*) and former guerillas to remove the mines littering the fields. Within a year, the province had restored more than 26,000 hectares of land, accounting for about half of all abandoned land and one-third of the province's agricultural land.[12] Within two years, the province reportedly restored to productive use 50,000 hectares of previously abandoned land.[13]

Because the authorities were able to mobilise large numbers of rural and urban people, good progress with land restoration was made in many areas. An example is Điện Bàn district. It had 114 subcommunes (villages), 93 of which were completely destroyed during the war. Many people faced hunger, and weeds had taken over their land. The district's new authorities mobilised everyone to turn 4,600 hectares of abandoned land into cultivated land. Former guerillas and local militia removed 20,794 landmines, during which 19 people were killed and 34 were injured.[14]

People from Hiền Lộc village in Bình Lãnh commune, Thăng Bình district, recalled that they returned after the war with 'two empty hands' (*với hai bàn tay trắng*). During the war, many working-age men had died, so many families returning to their villages were in a situation where 'a son had lost his father and a wife had lost her husband' (*cảnh con mất cha, vợ mất chồng*). Moreover, many families lacked the tools necessary for making a living and wild metre-high weeds had taken over their fields. Bombs and rockets had destroyed some of their land, and landmines remained in some rice fields. A widow with three children remembered:

---

12   Ibid.
13   Hội nghị tổ đổi công toàn tỉnh thành công tốt đẹp [The conference on labour exchange teams achieved good results], *Quảng Nam-Đà Nẵng*, 25 June 1977, p. 3.
14   Điện Bàn: Cả huyện là một công trường [Điện Bàn: The whole district is a working field], *Quảng Nam-Đà Nẵng*, 11 August 1976, p. 1.

> After the war, I took my three children home with two empty hands, no rice [*không lúa gạo*], no buffaloes. It was so miserable [*cực khổ lắm*]! This village was full of wild weeds and trees. We had to restore the abandoned fields by exchanging labour with others [*làm vần công với người khác*]. At that time we were afraid of mines exploding in the fields but we still tried to do [land restoration]. I was not afraid of death but worried that if I died, who would take care of the children? My sister-in-law died of a mine exploding when she was hoeing an abandoned field at that time.[15]

Villagers in Hiển Lộc recalled that the land tenure system had totally changed because farms had been abandoned for many years. The previous landlords had fled. Large areas of land now seemed to have a kind of common ownership. People restored any plot they liked as if it were their own. Some restored as much land as their families could manage. Those who came home first could select land close to their houses. Those who came later had to cultivate land further away.

While people restored some of the land on their own, the new authorities mobilised villagers to rehabilitate the remaining abandoned land. The new Thăng Bình district authorities mobilised villagers from less war-torn communes to help residents in heavily damaged communes. A former Bình Lãnh commune official recalled that people in Bình Nguyên, Bình Tú and Bình Trung communes who lived in or near the district centre came to help restore the fields in Bình Lãnh. After land restoration, the commune authority, through the local farmers' associations (*ban nông hội thôn*), reallocated land among households according to the number of people in their immediate family (*theo nhân khẩu*).[16]

The situation in Thanh Yên village in Bình Định commune (Thăng Bình district) was similar. After the war, returning residents found the place devastated, and weeds and bomb craters riddled their land. Local authorities mobilised people to restore abandoned land. Youth associations and former soldiers in the Saigon regime were mobilised to lead the campaign by restoring 'difficult' fields littered

---

15   Mr Đỗ in Hiển Lộc village recalled that he returned home from Đà Nẵng city later than other people so he was forced to cultivate land far from home that other people disliked (Author's interview, 14 October 2005, Bình Lãnh).
16   Ibid.

with landmines. Commune authorities and local farmers' associations reallocated all restored unclaimed land to households according to their needs.[17]

Land redistribution

After land restoration in QN-ĐN came land redistribution, for which—unlike in the Mekong Delta (see next sections)—there was no strong resistance from a landed class. At a conference to sum up the implementation of land policy and the LTTT (Land to the Tille) program in QN-ĐN on 30 July 1976, authorities announced that the province had 'successfully completed land redistribution to peasants':

> [O]ne year after starting to implement a new land policy, the fields in our province actually returned to peasants [*ruộng đất về tay nông dân*]. Basically, there is no more exploited class or landlords. Feudal exploitation has permanently been eliminated.[18]

According to authorities, the province had redistributed 19,547 hectares of arable land to 47,000 landless people. About 1,710 hectares had come from 'land donations' (*hiến điền*) and expropriations of land from landlords and 'lackeys of the imperialists' (*tay sai của Đế quốc*).

According to party researcher Lâm Quang Huyên, 'by May 1976, the former zone V [*khu V cũ*; in the Central Coast] had solved [its] land distribution problem'. Huyên reported that, according to the data from 61 communes and seven wards of nine districts in the Central Coast, the local authorities had appropriated 18,027 hectares, accounting for 31 per cent of the total arable land. This land was then allocated equitably to 34,875 land-poor and landless peasant households (containing 192,107 people).[19] The composition of appropriated land is displayed in Table 3.1.

17   Author's interviews, October–December 2005, Thanh Yên.
18   Ibid. At that time, the total agricultural land in the province was about 90,000 hectares.
19   Lâm Quang Huyên (*The Land Revolution in South Vietnam*, p. 180) also mentioned that Bình Trị Thiên province had retrieved 12,737 hectares of land and granted it to 40,609 peasant households.

Table 3.1 The composition of appropriated land in 61 communes
and seven wards in the Central Coast region

Type of land	Area (hectares)	Proportion (%)
Communal land (*công điền công thổ*)	4,515	25.05
Landlords' and rural capitalists' land	4,330	24.02
Rich peasants' land	1,717	9.52
Religious land	1,541	8.55
Other	5,924	32.86
Total	18,027	100.00

Source: Lâm Quang Huyên (1985), *Cách mạng ruộng đất ở Miền Nam Việt Nam* [*The Land Revolution in South Vietnam*], Hà Nội: NXB Khoa Học Xã Hội, p. 180.

In areas that were heavily war-damaged and where local authorities played a major role in restoring abandoned land, land redistribution was more extensive than in areas less affected by the war. For example, in Bình Lãnh and Bình Định communes in Thăng Bình district and Duy Phước commune in Duy Xuyên district, authorities redistributed equitably a large proportion of restored and communal land (*công điền*) to landless and land-poor households.[20] In addition, land redistribution happened gradually as families with more land lent some to their relatives and neighbours.[21] Meanwhile, in Hòa Tiến commune of Hòa Vang district, where levels of abandoned land and land restoration were more modest, local authorities granted only 'communal land' to peasants. They did not touch private land until collectivisation started, so inequitable land distribution remained.[22]

Other postwar economic restoration

With only 90,000 hectares of agricultural land—accounting for less than 10 per cent of natural areas—and a population of 1.5 million in 1976, QN-ĐN province had a low level of agricultural land per capita. In addition, most agricultural land was sandy and poor and

---

20   Xã Duy Phước trước bước ngoặc lịch sử [Duy Phước commune and its historic turning-point], *Quảng Nam-Đà Nẵng*, 24 September 1977, p. 1. One *sào* is equal to 500 square metres and 1 *thước* is equal to one-fifteenth of 1 *sào*.
21   Author's interviews, October–December 2005, Bình Định.
22   Hòa Tiến: 1,057 hộ tự nguyện đưa 379 ha ruộng đất vào làm ăn tập thể [Hòa Tiến: 1,057 households voluntarily put 379 hectares into collective farming], *Quảng Nam-Đà Nẵng*, 4 October 1977, p. 1.

had inadequate irrigation.[23] During the war, QN-ĐN's economy had depended heavily on imported commodities and food and foreign aid. The province produced only about 95,000–100,000 tonnes of food, falling short of its annual consumption of 450,000–500,000 tonnes.[24]

After the war, food security was the provincial leaders' main concern. To ensure self-sufficiency in food production, provincial leaders urged 'party members, cadres, and people to facilitate food production and economise to the maximum [*thực hành tiết kiệm tối đa*]'.[25]

In late 1976, the *Quảng Nam-Đà Nẵng* newspaper launched a column called 'The People's Forum' to discuss whether or not QN-ĐN province could resolve its own food problems. Several articles in this column came from state offices at provincial, district and commune levels. Most agreed that the province could feed itself. The methods for this included 'irrigation' (*thuỷ lợi*), 'intensive farming' (*thâm canh*), 'adopting new seeds' (*áp dụng giống mới*), 'developing subsidiary crops' (*phát triển cây màu*), 'increasing the number of crops per year' (*tăng vụ*), 'expanding agricultural land' (*mở rộng diện tích*) and 'transforming and designing fields' (*cải tạo đồng ruộng*).[26] Provincial leaders eventually asserted that the province could ensure its own food supply and they set a target to produce 500,000 tonnes of staple food in 1980. The plan called for an expansion of agricultural land, from 50,000 hectares to 140,000 hectares, extending irrigated areas from 21,000 hectares in 1976 to 60,000 in 1980, moving 160,000 people from lowland areas to new

---

23   Phấn đấu mở rộng nhanh diện tích canh tác [Strive to extend cultivated area], *Quảng Nam-Đà Nẵng*, 26 June 1976, p. 4.

24   CTKQN, *Quảng Nam's Socioeconomic Development*, p. 22.

25   Đẩy mạnh sản xuất và thực hành tiết kiệm giải quyết vấn đề lương thực cấp bách trước mắt [Increase production and be thrifty to immediately deal with urgent food shortage], *Quảng Nam-Đà Nẵng*, 16 February 1976, p. 1; Nêu cao tinh thần tự lực tự cường trong sản xuất và xây dựng quê hương [Be self-reliant in ensuring food production and building the country], *Quảng Nam-Đà Nẵng*, 29 September 1976, p. 1.

26   Tỉnh ta có khả năng tự giải quyết lương thực hay không? [Is our province able to solve our own food problem?], *Quảng Nam-Đà Nẵng*, 22 November 1976, p. 1; Nhìn lại diện tích đất đai để thấy rõ khả năng tự giải quyết lương thực [Re-examining agricultural areas to evaluate our capacity for dealing with food problems], *Quảng Nam-Đà Nẵng*, 26, November 1976, p. 1; Nước và sản xuất lương thực ở tỉnh ta [Irrigation and food production in our province], *Quảng Nam-Đà Nẵng*, 18 December 1976, p. 1; Vì sao tỉnh ta đặt vấn đề giải quyết lương? [Why do we pay great attention to solving the food production problem?], *Quảng Nam-Đà Nẵng*, 22 December 1976, p. 1.

economic zones in mountainous areas, increasing subsidiary crops to 30 per cent of total food production and expanding the area of new spring–summer rice crops to 15,000 hectares.[27]

QN-ĐN's leaders considered irrigation an important first measure (*biện pháp hàng đầu*). In November 1975, they launched a widespread campaign to expand irrigation, mobilising people to dig ponds, build dams and canals and use manual pumps to water fields.[28] Within the first three months of 1976, QN-ĐN had mobilised the equivalent of 111,850 days of labour to repair 363 dams and 132 canals totalling 131,905 metres. In addition, the province started five large-scale irrigation projects, including the 'Phú Ninh great irrigation dam' (*Đại công trình thuỷ lợi Phú Ninh*) in Tam Kỳ district and Trường Giang and Cao Ngạn dams in Thăng Bình district.[29]

Villagers in Hiền Lộc and Thanh Yên villages recalled that, soon after reunification, each family contributed months of labour to build Cao Ngạn dam in Bình Lãnh commune and Phước Hà dam in Bình Phú commune. Later, authorities mobilised village youth to build another great irrigation dam, Phú Ninh, in Tam Kỳ district. One woman in Hiền Lộc village told me she had to contribute three months of labour to the construction of Cao Ngạn dam and many days for the other dams. When undertaking this work, she had to carry her own food.[30] Some peasants in Hiền Lộc said governments in the French and American times had attempted to build these dams but were unable to pay for labour and the use of private land. Under the revolutionary authorities, however, land belonged to everyone (*của chung*) so it was easier to build dams, roads and other infrastructure that required vast tracts of land.[31]

---

27   Hồ Nghinh: Quảng Nam-Đà Nẵng vượt bậc phát triển sản xuất nông nghiệp [Hồ Nghinh: Quảng Nam-Đà Nẵng has made great progress in agriculture], *Nhân Dân*, 8 March 1977, p. 5; Nghị quyết hội nghị Ban chấp hành Đảng bộ tỉnh khóa 11 [Resolution of 11th Provincial Party Executive Committee], *Quảng Nam-Đà Nẵng*, 12 March 1977, p. 1.

28   Đẩy mạnh công tác thủy lợi nhỏ để phục vụ sản xuất xuân hè và hè thu [Extending irrigation for the spring–summer and summer–autumn crops], *Quảng Nam-Đà Nẵng*, 8 March 1976, p. 1. Bình Dương commune in Thăng Bình district was considered an exemplary case because it had dug 2,200 ponds (*giếng*) from which to water its crops. On average, each labourer had dug one pond.

29   Toàn tỉnh sôi nổi ra quân làm thủy lợi lợi [People in the province are extending irrigation], *Quảng Nam-Đà Nẵng*, 12 May 1976, p. 1.

30   Author's interview, 15 October 2005, Bình Lãnh.

31   An elderly man in Hiền Lộc village referred to Phú Ninh dam as *Ba kỳ* dam ('the three periods dam') because it was initiated by the French, continued by Saigon's government and completed by the new government (Author's interview, 17 October 2005, Hiền Lộc).

Dams took over large amounts of peasants' land, but I found no evidence of strong resistance, although some peasants did express dissatisfaction with the policy. An article in *Quảng Nam-Đà Nẵng* newspaper (on 29 September 1976) told how the party cell of Hòa Nhơn commune in Hòa Vang district overcame peasants' 'backward thoughts and superstitions' when it decided to open a canal through hills and villages to divert water from the river to rice fields. Many peasants refused to participate in the project. Some elderly people were afraid the dam would 'break down the heart of their village land' (*đứt con đất của làng*) and upset 'the spirits of the land' (*Thổ địa quở phạt*). Some worried about the loss of their land and their family's tombs. Other residents doubted the success of the project. To overcome these objections, the party cell organised meetings to 'fight and criticise feudal thoughts such as selfishness and superstitions'.[32]

As well as more irrigation, local authorities wanted new rice seeds used and more crops per year. To solve food shortages in the interval between the winter–spring and summer–autumn crops (*chống đói giáp hạt*), the province launched a campaign to adopt a new spring–summer rice crop (*vụ xuân hè*).[33] Such a crop was new to many peasants in QN-ĐN who had previously cultivated at most only two rice crops per year. However, adopting the spring–summer crop achieved good initial results.[34] Some years later, many of the irrigated areas of QN-ĐN also cultivated a third rice crop each year.

Peasants who had been under the influence of the Saigon government's rural development program were familiar with the adoption of new rice seeds. For example, QN-ĐN peasants in Điện Quang commune in Đại Lộc district, Điện Minh and Điện Phương in Điện Bàn district and Hòa Nam and Hòa Xuân communes in Hoà Vang district had used chemical fertilisers, new rice seeds and farm machinery even before 1975.[35] However, the majority of peasants in the province still grew traditional rice varieties and had rarely used chemical fertilisers or human manure (*phân bắc*: literally, 'northern manure').

---

32 Be self-reliant in ensuring food production, *Quảng Nam-Đà Nẵng*, 29 September 1976, p. 4.
33 Tăng vụ sản xuất xuân hè [New additional spring–summer crops], *Quảng Nam-Đà Nẵng*, 16 February 1976, p. 4.
34 Vụ sản xuất xuân hè thắng lợi [The spring–summer crops have a good result], *Quảng Nam-Đà Nẵng*, 7 August 1976, p. 1.
35 Những mùa lúa đầu tiên [The first rice crops], *Quảng Nam-Đà Nẵng*, 19 April 1976, p. 2.

As an additional step towards increasing rice production, authorities in QN-ĐN called for the removal of tombs from agricultural land.[36] Although this policy touched a sensitive aspect of peasant culture—which considered ancestral tombs immovable—it encountered only modest resistance. For example, people in Điện Bàn district 'spent 100,000 working days to remove 90,000 scattered tombs, extending 80 additional hectares of agricultural land'.[37] Despite dissatisfaction with the policy, few Điện Bàn peasants openly objected. Many, though, criticised it in private. Similarly, in Duy An commune, Duy Xuyên district, peasants mockingly said 'even the dead aren't allowed to rest' (*người chết cũng không được nằm yên*). However, authorities were finally able to 'convince' these peasants to accept the policy.[38]

According to the *Quảng Nam-Đà Nẵng* newspaper, postwar economic restoration policies achieved good results. From mid-1975 to the end of 1977, the province expanded its cultivated areas (*diện tích gieo trồng*) from 96,000 to 183,337 hectares, equal to the figure in 1965. Production of staple food also increased, from 149,062 tonnes in 1975 to 300,000 tonnes by the end of 1977, showing the province would be able to overcome food shortages and produce its target of 500,000 tonnes by 1980.[39]

Building the foundation for collective farming

While carrying out postwar economic restoration policies, authorities also created labour exchange teams (*tổ đổi công vần công*). This policy seemed to fit well with local practices in which reciprocity and mutual assistance were still popular among villagers. Also, before 1975, and especially during the Việt Minh period, revolutionary authorities in many areas of QN-ĐN had organised peasants into labour exchange teams and even some collectives.[40]

---

36   Ủy ban nhân dân ra chỉ thị về công tác quy hoạch mồ mả và nhà của của nhân dân [The Provincial People's Committee issued a directive to reallocate tombs and houses], *Quảng Nam-Đà Nẵng*, 28 August 1976, p. 1.

37   Điện Bàn: The whole district is a working field, *Quảng Nam-Đà Nẵng*, 11 August 1976, p. 1.

38   Duy An lập khu nghĩa địa mới [Duy An has established new graveyards], *Quảng Nam-Đà Nẵng*, 5 April 1976, p. 2.

39   Gióng đường cày thắng lợi [Be victorious in agriculture], *Quảng Nam-Đà Nẵng*, 26 April 1977, p. 1.

40   Ngành nông nghiệp tỉnh Quảng Đà tích cực chăm lo vụ mùa tháng 8 [Agricultural sector in Quảng Đà is positive about caring for August crops], *Quảng Đà*, 20 June 1974, p. 1.

Soon after the war, Điện Bàn district, for example, formed 598 labour exchange teams to help with land restoration.[41] Likewise, peasants in Hiến Lộc and Thanh Yên villages, Thăng Bình district, recalled that, to undertake land restoration, subcommune farmers' associations organised them into labour exchange teams, each comprising 15–20 neighbouring households.[42] Most of these labour exchange teams operated in an irregular and seasonal manner (*tổ đổi công không thường xuyên, thời vụ*) and for specific tasks such as preparing fields and harvesting. They were dismantled when the specific task was completed.

Local authorities successfully organised peasants into 'regular labour exchange teams' (*tổ đổi công thường xuyên*) in some parts of QN-ĐN. For example, Sông Bình subcommune (Đại Quang commune, Đại Lộc district) formed regular labour exchange teams for land restoration, production and irrigation. Some 160 households in the subcommune were organised into such teams, each comprising 12–14 households and one or two buffaloes. Members of these organisations exchanged labour among themselves in their everyday production activities. Men were often in charge of hoeing and ploughing, while women did lighter work such as transplanting and harvesting. Those who did not have draught animals could use the team's buffaloes.[43] Another example is Thăng Phước commune in Thăng Bình district, whose land had been abandoned for 10 years. Most of its men died in the war, and women and the elderly made up its workforce. To cope with such difficulties, local authorities quickly organised peasant households into 39 regular labour exchange teams.[44]

At a conference of labour exchange teams in June 1977, QN-ĐN leaders (*hội nghị tổ đổi công*) praised the role of labour exchange in 'training peasants to work collectively' and solving their postwar problems. Leaders criticised these organisations, however, for 'developing unevenly and unsoundly' and operating according to simple, unfair and irrational methods (*chưa công bằng, hợp lý*). Therefore, they called for the upgrading of simple labour exchange teams into a higher-level organisation called a 'production team working according to norms

---

41 Điện Bàn: The whole district is a working field, *Quảng Nam-Đà Nẵng*, 11 August 1976, p. 1.
42 Author's interviews, October–December 2005, Thanh Yên and Hiến Lộc.
43 Tổ đổi công vần công ở Sông Bình [Labour exchange teams in Sông Bình], *Quảng Nam-Đà Nẵng*, 8 May 1975, p. 2.
44 The whole population of Thăng Phước commune, *Quảng Nam-Đà Nẵng*, 23 May 1977, p. 2.

and contracts' (PTWCNC) (*tổ sản xuất có định mức, khoán việc*).[45] According to the guidelines, a PTWCNC was based on individual ownership of land and other means of production, but management was similar to that in a collective organisation. For example, officials kept track of labour exchanges through work-points, norms and contracts, and they distributed state agricultural inputs to each team.[46]

Establishing the PTWCNCs was the 'first step of collectivisation', and the aim was to 'facilitate peasants' solidarity', 'improve collective work', 'establish state and peasant relations', 'make peasants familiar with collective work' and 'select and train cadres' for ongoing collectivisation.[47] However, unlike with a simple labour exchange team, building a PTWCNC required training cadres and peasants.

By October 1977, QN-ĐN had trained nearly 9,000 cadres; some districts had completed the training of all cadres and were preparing to establish PTWCNCs before the winter–spring crop of 1977–78.[48] However, there were also some difficulties in building PTWCNCs. For example, Đại Lộc district selected Đức Phú subcommune in Đại Hiệp commune in which to build pilot PTWCNCs. To ensure success, the authorities had to do a lot of preparation, such as organising policy, studying training (*học tập chính sách*) for cadres and peasants and preventing peasants from selling draught buffaloes and farm tools. Authorities were able to mobilise 97 per cent of households into 33 PTWCNCs, but these organisations did not function well. Cadres were confused about management (*lúng túng về quản lý*) and unsure how to make norms and contracts and calculate and determine workdays among households.[49] Similarly, An Bình subcommune (in Tiên Kỳ commune, Tiên Phước district) faced difficulties managing its PTWCNCs. Peasants were

45   The conference on labour exchange teams, *Quảng Nam-Đà Nẵng*, 25 June 1977, p. 1.

46   Xây dựng các tổ sản xuất có định mức khoán việc [Establishing production teams working according to norms and contracts], *Quảng Nam-Đà Nẵng*, 22 October 1977, p. 3.

47   Bản hướng dẫn nội dung xây dựng tổ đổi công có định mức, khoán việc [Guidelines for establishing production teams working according to norms and contracts], *Quảng Nam-Đà Nẵng*, 29 June 1977, p. 1.

48   Establishing production teams working according to norms and contracts, *Quảng Nam-Đà Nẵng*, 22 October 1977, p. 3.

49   Đại Lộc xây dựng các tổ sản xuất có định mức khoán việc [Đại Lộc is establishing production teams working according to norms and contracts], *Quảng Nam-Đà Nẵng*, 26 October 1977, p. 2. In June 1977, the provincial authorities released a directive forbidding merchants from purchasing and slaughtering draught animals (Nghiêm cấm thương nhân mua trâu bò để giết thịt [Prohibiting private merchants from purchasing and slaughtering livestock], *Quảng Nam-Đà Nẵng*, 24 September 1977, p. 1).

confused about how to work according to norms and contracts. Some complained that the procedures 'coerced, humiliated and restricted' people and 'did not raise their enthusiasm'. One peasant complained that, 'without norms and contracts, I can work with all my heart. Now under the norms and contracts, I do enough to just achieve satisfactory results according to the contract!'[50]

By the end of 1977, when collectivisation began, QN-ĐN had built 4,524 simple interim collective organisations, of which nearly 80 per cent of peasant households were members. Of these, 2,625 were PTWCNCs, although 42 per cent of them were below standard.[51] Among the places in QN-ĐN without any PTWCNCs were Hiển Lộc village (Bình Lãnh commune) and Thanh Yên village (Bình Định commune).[52]

Generally speaking, by late 1977, authorities in QN-ĐN, particularly in Thăng Bình district, were able to accomplish most of the intended preparatory measures for collectivised farming. The story is different in the Mekong Delta's An Giang province.

Preparing for collectivisation in the Mekong Delta

Consolidating local authorities

After the war, An Giang province and other parts of the Mekong Delta were under the control of the new military administration (*thời kỳ quân quản*). It took a year for the new authorities to consolidate a civilian government in An Giang and other provinces.

According to the documents relating to Chợ Mới district's party committee, after the war, the new district authorities faced many difficulties in controlling society and consolidating their power.[53]

---

50 Bình An xây dựng tổ sản xuất có định mức khoán việc [Bình An is establishing production teams working according to norms and contracts], *Quảng Nam-Đà Nẵng*, 24 April 1978, p. 3.

51 Tỉnh ủy mở hội nghị bàn về cải tạo xã hội chủ nghĩa đối với nông nghiệp [Provincial Party Committee opens conference on socialist agricultural transformation], *Quảng Nam-Đà Nẵng*, 22 February 1978, p. 1.

52 Author's interviews, August–November 2005, Hiển Lộc and Thanh Yên villages.

53 Đảng Bộ Chợ Mới (2000), Trên mặt trận bảo vệ an ninh tổ quốc [On the national security front], in *Chợ Mới 25 năm xây dựng và phát triển* [*Chợ Mới's Socioeconomic Development over the Past 25 Years*], Chợ Mới: Đảng Bộ huyện Chợ Mới, p. 44.

According to a former official of Long Điền B commune, after 30 April 1975, Bảo An soldiers (of the Hòa Hảo religion) and soldiers and officers from Saigon gathered in Chợ Mới district and fought against the revolutionary force for a week.[54] Assessing the difficulties of Chợ Mới district in the first few years after reunification, a former party secretary there reported that

> eighty per cent of the population was religious; most of them were the Hòa Hảo. Twenty thousand Saigon soldiers gathered here ... forty per cent of the population were landless and land-poor.[55]

The authorities considered the large number of former Saigon and Bảo An soldiers in Chợ Mới a political and military threat.

Another difficulty authorities faced was a lack of local party cadres to fill new positions. This was the situation in Chợ Mới district and many other parts of the Mekong Delta.[56] During the war, local networks of southern cadres had been destroyed and many revolutionaries killed, especially through the American and Government of South Vietnam Phoenix program.[57] After the war, party organisations in An Giang were weak, and 17 communes had no party cells. Most surviving ex-revolutionaries came from remote districts such as Tịnh Biên, Tri Tôn and Phú Châu.[58] Villagers in Chợ Mới called their area a 'white area' (vùng trắng), which meant no Communist Party cells operated there until reunification. By mid-1975, Chợ Mới district had only 58 Communist Party cadres—insufficient for establishing a new local authority. Therefore, 40 party cadres were sent from nearby Sa Dec province.[59] In assessing the situation of the party organisation in Chợ Mới district in the first few years after reunification, the secretary

---

54 Author's interview, 16 June 2005, Chợ Mới.
55 Qua hội nghị công báo hoàn thành cơ bản hợp tác hóa nông nghiệp ở Chợ Mới: Bài học gì được rút ra [Report from a conference announcing the completion of collectivisation in Chợ Mới: Lessons learned], An Giang, 15 April 1985, p. 1.
56 Phạm Văn Kiết, Peasants are eager for collective farming, p. 28.
57 Beresford, M. (1988), Issues in economic unification: Overcoming the legacy of separation, in D. Marr and C. White (eds), Postwar Vietnam: Dilemmas in Socialist Development, Ithaca, NY: Cornell University Press, p. 107; Hicks, Organizational adventures in district government, p. 120.
58 Ủy Ban Nhân Dân Tỉnh An Giang [hereinafter UBNDTAG] (2003), Địa Chí An Giang [An Giang Province], An Giang: Ủy Ban Nhân Dân Tỉnh An Giang, p. 349.
59 Ban Chấp Hành Đảng Bộ Huyện Chợ Mới [hereinafter BCHDBHCM] (1995), Lịch sử Đảng bộ huyện Chợ Mới [The History of Chợ Mới Party Cell, 1927–1995], Chợ Mới: Ban Chấp Hành Đảng Bộ Huyện Chợ Mới, p. 169.

of An Giang commented that 'the party bases [*cơ sở đảng*] were small and thin [*mỏng*]. Mass organisations and communal and hamlet authorities were inadequate and weak.'[60]

In the first few years after reunification, many communes in Chợ Mới district had no party cells or only weak ones. For example, in 1977, Long Điền B commune had a new party cell with only three members—one was the secretary of the cell, one was the commune's chairman (*chủ tịch xã*) and the other was the commune's chief police officer (*trưởng công an xã*), who had just become a party member. At the hamlet level, new authorities selected some trusted local people to work as chiefs (*trưởng ấp*) and members of managerial boards and peasant associations.

A majority of local cadres in Long Điền B commune were not ex-revolutionaries. They were selected thanks mostly to the revolutionary merit of their parents, brothers or even distant relatives. Long Điền B residents remembered after reunification new local cadres often called themselves 'ex-undercover revolutionaries' (*cán bộ nằm vùng*) or 'meritorious for having hidden Vietnamese communist cadres' (*có công nuôi cán bộ*). In many cases, the new cadres were exaggerating. One elderly man, a former chief of the Saigon government's local militia group (*dân quân tự vệ*) in a hamlet in Long Điền B, commented that the local chief had been a member of his staff during the war. After reunification, the man was made hamlet chief thanks to the revolutionary merit of his brother-in-law, who lived in Đồng Tháp. The man often claimed that he had previously been an undercover revolutionary, but many people did not believe him.[61] Some local people added that, due to lack of revolutionary merit, many of these cadres tended to work to gain political merit (*lập công*). Some said that, while these cadres tried to comply with official policies, they also pursued their own interests.

Building the foundation for collective farming

While establishing a new government, leaders in the Mekong Delta also began to build the foundations for collectivisation. In the Central Coast region, the first stage had been to create labour exchange teams. In the Mekong Delta, however, the first step was to create 'production

60   Huyện Chợ Mới hoàn thành hợp tác hóa nông nghiệp [Chợ Mới district has completed collectivisation], *An Giang*. 4 April 1985, p. 1.
61   Author's interview, 29 June 2005, Long Điền B, Chợ Mới.

solidarity teams' (*tổ đoàn kết sản xuất*). The different names reflect social and economic differences between the two regions. In the Mekong Delta, labour exchange among peasants had not been as common in previous decades as in the Central Coast. Instead, in the Mekong Delta, land-rich peasants had often hired the land-poor and landless to work for them. The term 'solidarity' reflected the Communist Party government's desire to unite these two classes of villagers.

According to official guidelines, each production solidarity team should farm 30–50 hectares. The authorities expected that, in so doing, cadres and peasants would learn to exchange labour (*vần đổi công*) and work together collectively. In practice, many production solidarity teams did not operate this way. Villagers in Long Điền B said the production solidarity teams in Chợ Mới district often had 200 to 300 hectares—virtually the size of a hamlet. Moreover, many production solidarity teams did not exchange labour. A former production solidarity team leader said peasants refused to farm that way, wanting to hire labour as they had previously, rather than exchanging it, because the latter method was unknown and considered inferior.[62] Therefore, although production solidarity teams existed, people farmed no differently to before—as individuals, not as teams. One local cadre described this as 'each person cultivated his own land and paid his own fees' (*đất ai nấy làm phí ai nấy trả*). The production solidarity teams played only an intermediary role between the state and peasants. The teams were in charge of delivering agricultural inputs and other necessary goods from the state to peasants and collected taxes from peasants for the state.[63]

In 1976, Chợ Mới district established 105 production solidarity teams in 101 hamlets, but their quality was low. Chợ Mới's party cell reported:

> In 1977, the district party committee realised that in reality peasants in these organisations still farmed individually. In other words, these organisations were in fact just fuel-delivering teams [*tổ xăng dầu*]. Therefore, the district's party committee decided to establish a committee for agricultural transformation [*ban cải tạo nông nghiệp huyện*] and immediately selected several cadres to go to the provincial capital for training.[64]

---

62   Author's interview, 25 June 2005, Long Điền B.
63   Author's interview, 15 June 2005, Long Điền B.
64   Report from a conference, *An Giang*, 15 April 1985, p. 1.

By 1978 An Giang province had established about 300 production solidarity teams, 528 by 1979 and 1,528 by 1980. Most of these organisations, according to assessment reports, were 'inadequate in quality and scale' (*không đúng tính chất và quy mô*).[65] Hoping to improve their quality, the vice-chief of An Giang's committee for agricultural transformation in June 1981 ordered the teams to be made smaller.[66] However, the situation did not improve much because local leaders were preoccupied with land redistribution and other issues.

Prohibiting non-resident cultivators (*cắt xâm canh*)

Compared with other regions of the south, the Mekong Delta (especially the western part, Miền Tây, where An Giang is) was among the least affected by the war. Thanks to a relatively peaceful life, abundant natural resources and previous agrarian reforms, food production in many parts of the region had exceeded consumption needs. Previous reforms, especially the LTTT program, had almost eliminated big landlords, establishing a solid institutional foundation of small and family-owned farms;[67] and the rural population had diverse occupations, including growing commodity crops, working as labourers, engaging in petty trade and other non-farming work. This made the social structure and economic activities of the Mekong Delta more diverse than in the Central Coast and other regions of the south.

Another feature of difference in the region was that peasants' farming and production activities extended beyond their villages. Many peasants had land in their own hamlet as well as in distant communes, districts and even other provinces. In Chợ Mới district (An Giang), more than half the peasants also had fields elsewhere. Most of their 'outside' land was in Long Xuyên quadrangle (*Tứ Giác Long Xuyên*) and the Plain of Reeds (*Đồng Tháp Mười*), which local peasants called 'big fields' (*đồng lớn*). A former district official mentioned that the combined area of agricultural land Chợ Mới's peasants had outside the district exceeded the district's total agricultural land. Most peasants who had

---

65   Sở Thông Tin Văn Hóa An Giang [hereinafter STTVHAG] (1978), *Thông tin phổ thông* [*General Information*], Vol. 9, Long Xuyên: NXB Sở Thông Tin Văn Hóa An Giang, p. 9; Phong trào hợp tác hóa tiếp tục đi vào chiều hướng ổn định và phát triển theo hướng phương châm tích cực và vững chắc [Collectivisation continues to progress positively and firmly], *An Giang*, 7 June 1981, p. 2.

66   Collectivisation continues to progress positively and firmly, *An Giang*, 7 June 1981, p. 2.

67   Callison, *Land-to-the-Tiller in the Mekong Delta*, p. 328.

land in 'big fields' possessed more than 100 *công* (10 hectares) of land; despite growing one 'floating rice' crop (*lúa nổi*) a year, these peasants annually carried home thousands of *gịa* of paddy.[68] Many peasants in Long Điền B recalled that, before 1975, farming was a profitable job and they enjoyed a high standard of living. Even agricultural labourers who did not have land earned a 'sufficient' livelihood (*sống thoải mái*) working for the land-rich or in fishing and non-farming occupations.[69]

As in many other authoritarian states, Vietnam's new government attempted to remake the complex rural south so it would be easy to manage the population and production system in line with its existing socialist model of administration.[70] To deal with the complexity of the rural south and restrict landed peasants from cultivating beyond their own villages, officials in Hanoi issued Directive No. 235/CT-TW in September 1976:

> For land of non-commune residents [*đối với ruộng đất xã này xâm canh xã khác*], if it belongs to labouring peasants [such as poor and middle peasants], let them continue to cultivate it; if the land has been classified as land under confiscation [such as the land of rich peasants and landlords], then grant it first to current cultivators of the land who have farmed the land for a long time and now do not have enough land.[71]

The Politburo's Directive No. 57 (15 November 1978) clarified that land confiscation would apply to the land of non-resident rich peasants, rural capitalists and upper–middle peasants. For the land of non-resident labouring people, local authorities should either mobilise landowners to the site of their land or give them other land (in their village) in exchange.[72] In light of these directives, the authorities in An Giang and elsewhere in the Mekong Delta implemented a policy of prohibiting 'land occupying' or 'non-resident cultivators' (*cắt xâm canh*), which meant many peasants in the region, including middle peasants, had to give up much of their 'outside' land.[73]

---

68 Author's interview, 3 August 2005, Long Điền B. One *công* is equal to one-tenth of 1 hectare. One *gịa* is equal to 20 kilograms.
69 Author's interview, 17 August 2005, Long Điền B.
70 Scott, *Seeing Like a State*.
71 ĐCSVN, Secretariat's Directive No. 273, pp. 280–1.
72 ĐCSVN (2004), Chỉ thị 57/CT-TW về việc xóa bỏ các hình thức bóc lột của phú nông, tư sản nông thôn và tàn dư bóc lột phong kiến [Directive No. 57 on eliminating exploitation in the south], in ĐCSVN, *Văn Kiện Đảng Toàn Tập, Tập 38, 1977* [*Party Document: Volume 38, 1977*], Hà Nội: NXB Chính Trị Quốc Gia, p. 474.
73 Huỳnh Thị Gấm, Socioeconomic changes in the Mekong Delta, p. 88.

Villagers in Long Điền B commune, Chợ Mới district, recalled that this was one of the most controversial postwar policies they faced. It encouraged 'cultivation close to the residential area' (*liền canh liền cư*) and people were not allowed to farm outside their residential communes or districts.[74] According to some former local officials in Chợ Mới district, this policy enabled the new authority to control rural society and food production and procurement. They also argued that, if people were allowed to move freely, the local authorities would not be able to mobilise people into collective organisations. Moreover, these policies were a first step towards land redistribution and collectivisation. Prohibiting non-resident cultivators helped authorities appropriate the land of non-resident land-rich households and give it to land-poor and landless households in each commune.[75] In addition, the policy also helped to identify land for state farms (*nông trường*), district farms (*nông trang*) and other state organisations.

Under the non-resident cultivator prohibition, many peasants in Chợ Mới claimed to have lost land, in Tứ Giác Long Xuyên and Đồng Tháp Mười, to newly established state farms, collectives or production units. Many expressed their dissatisfaction. Some peasants resisted quietly by abandoning land and refusing any other land in exchange.

Mr Ph, a farmer in Long Điền B commune, Chợ Mới district, recalled that he had owned 7 hectares of land in another district of An Giang, Châu Thành, since 1952. After 1975, he continued to till the land for two years until a state farm was established and he was expelled. He said he felt sad to lose the land, but could not do anything about it. He returned home to borrow 5 *công* (0.5 hectares) of land from his relatives from which to make a living.[76]

Mr H, who lived in the same hamlet as Mr Ph, had possessed 20 hectares of land in Thoại Sơn district of An Giang since 1954. He said a state farm appropriated his land and offered him other land in exchange, but he was very upset and refused the offered land. Eventually, he decided to abandon the land and returned home to 'raise ducks and chickens and work on a few *công* of land around the house'.[77]

---

74   Author's interviews, August–October 2005, Long Điền B.
75   Author's interviews, 20–29 June 2005, Long Điền B.
76   Author's interview, 3 August 2005, Long Điền B.
77   Author's interview, 4 August 2005, Long Điền B.

In general, peasants were submissive and kept their true feelings private. Most claimed they feared the new authorities because they did not yet know the laws properly and, even if they tried their best, those in power would control the situation.

Some peasants, however, reacted strongly and openly confronted officials. For example, Mr Ba G in Long Điền B, who lost more than 10 hectares of land in Thanh Bình district of Đồng Tháp province one year after reunification, complained to the commune chairman:

> In my opinion, what you did was unconscionable. In the past, we endured the war with you to cultivate on the land, as you know. We used to have meals and drink with you. We lived with you for many years, suffering a lot from wars. Many people had died in this place during that time. Now you say we usurped the land and you expel us. Pity us![78]

He cried for hours before the chairman, but the chairman remained silent, doing nothing to help him. Finally, knowing that it was not possible to change the situation, Mr Ba G decided to abandon his land and offered a mocking goodbye to the official:

> [Y]ou often said that you liberated us from the yoke of slavery, but now you put us with another yoke, the yoke of no land with which to make a living![79]

A woman in Long Điền B who lost 13 hectares of land in another An Giang district shared her story:

> After reunification, the authorities planned to establish a state farm on the peasant land. We complained to the commune authorities. In response to our objection, the commune secretary organised a meeting with us. She, the commune secretary, suggested exchanging our land. I got angry and said, 'Whose land? Peasants' land or your land? If it is your land, we will take it but other people's land we refuse.' I turned back and said to the crowd: 'Those who do not want to exchange land, raise your hands.' They all raised their hands ... a man from the crowd stood up and said, 'I come from communist areas but I haven't seen anyone like you. Now we do not have any land to till.' The commune secretary could not do anything but withdrew in silence ... Then, our small group, about 20 people, went to the district authorities to complain. They said they would consider our petition

---

78 Author's interview, 30 June 2005, Long Điền B.
79 Ibid.

and would resolve it gradually. But nothing had been resolved until Mr Nguyễn Văn Linh ascended to power. As a result, I had to go around and work as a wage-earner for seven to eight years.[80]

This 'prohibition' was controversial not only among peasants in Long Điền B, but also with many other peasants in the Mekong Delta. They argued that peasants in the delta had always enjoyed 'freedom' over where they resided, in selecting their occupations and in seeking new economic opportunities. Many peasants in Long Điền B commune recalled that people became sick and died due to depression after losing their land. A former cadre of a production unit in Long Điền B said:

> Southern people valued their land highly: 'First are children, second is land [nhất hậu hôn nhì điền thổ].' Because many suffered a lot to accumulate the land during the wars, when losing their land, they were so sad that a few of them suffered mental sickness, even died of mental depression.[81]

Villagers considered the prohibition an 'odd' policy (chính sách kỳ cục) imposed from the north. Land occupation meant occupying the land of others (xâm lấn), but in reality people put great effort into reclaiming and improving their land rather than taking land belonging to others.[82] Mr H. H. in Long Điền B recalled losing 2 hectares of land in Long Điền A, a nearby commune, and a commune official began working on his land. Mr H. H. said:

> They [local officials] told me that I was not allowed to cultivate the land because I was not a commune resident. One had to cultivate where one lived. I argued that now the north, the centre and the south were united into one country, people had the right to cultivate anywhere. I did not steal anybody else's land!

He failed to convince the officials, although he was later granted a few công of redistributed land when he joined a production unit in Long Điền B.[83]

---

80 Author's interview, 10 August 2005, Long Điền B. Nguyễn Văn Linh was the VCP's general secretary from 1987 until 1992. Local people often divided the period 1975–90 into two phases: Lê Duẩn's phase (trào Lê Duẩn) and Nguyễn Văn Linh's phase (trào Nguyễn Văn Linh).
81 Author's interview, 20 September 2004, Long Điền B.
82 Author's interview, 4 August 2005, Long Điền B.
83 Author's interview, 9 August 2005, Long Điền B.

A report of the Communist Party's Central Committee for Agricultural Transformation in 1984 recognised the shortcomings of, and variation from one area to another in the implementation of, the prohibition policy. Some local authorities implemented the policy correctly by encouraging peasants who had land in two different areas (*hai nơi*) to choose only one. But in general, many local authorities were often confused about how to resolve the problem. Some considered non-residents' land 'invaded' (*coi xâm canh là xâm lấn*). They implemented the policy incorrectly, and non-resident peasants were often coerced into abandoning their land, even though they had not been given land in exchange in their residential area.[84]

First land redistribution

The VCP's Resolution No. 24, Directive No. 253/NQ-TW (20 September 1976) and Directive No. 28/CT-TW (26 December 1977) called for land redistribution with the aim of 'eliminating vestiges of feudalism' and the 'exploitation of land'. In response, An Giang and other provinces in the Mekong Delta implemented these policies but often interpreted them to mean 'mobilising land donations' (*vận động hiến điền*) and redistributing land among peasants in the spirit of 'sharing one's rice and clothes' (*nhường cơm sẻ áo*).[85] A former member of the Provincial Committee for Agricultural Transformation remembered emphasising the 'sharing of one's rice and clothes' rather than 'eliminating exploitation' during reform in An Giang from 1976 to 1980. The provincial authorities also paid great attention to the economic objectives of the reform.[86]

According to a Chợ Mới party cell report, during the period of military control (May 1975 to February 1976), the district 'confiscated 2,214 hectares of land abandoned by reactionaries who had left Vietnam[87] and temporarily granted land to 3,760 landless and land-poor households'.[88]

---

84 BCTNNMN, *Report on Land Redistribution*, p. 17.
85 These terms can be found in Tài liệu hỏi về hợp tác xã nông nghiệp [Inquiry about collective farming] in STTVHAG, *General Information*, p. 9. Ban Cải Tạo Nông Nghiệp An Giang (1978), *Báo cáo tình hình cải tạo xã hội chủ nghĩa* [*Report on Socialist Agricultural Transformation*], 13 December, Long Xuyên: Ban Cải Tạo Nông Nghiệp An Giang; Công tác điều chỉnh ruộng đất ở quê nhà [Land redistribution in rural areas], *An Giang*, 6 September 1982, p. 1; Trích diễn văn của đồng chí Lê Văn Nhung [An extract from Lê Văn Nhung's speech], *An Giang*, 10 May 1985, p. 1.
86 Author's interview, 31 May 2005, Long Xuyên.
87 Many of them were 'boat people' who fled Vietnam by sea after the war.
88 BCHDBHCM, *The History of Chợ Mới Party Cell*, p. 173.

Afterwards district authorities emphasised 'land redistribution' (*chia cấp ruộng đất*) in the spirit of 'sharing one's clothes and rice' among the peasants. However, the authorities persistently pressed land-rich households to share any land in excess of their farming capacity with landless and land-poor households.[89]

People in Long Điền B remembered the land policy as being one of 'sharing one's rice and clothes' (*chia cơm sẻ áo*) or 'land sharing' (*trang trải ruộng đất*). A former commune official who had been in charge of Long Điền B's agricultural transformation said the first post-1975 land policy he carried out in his commune was 'land redistribution' (*điều chỉnh ruộng đất*). In his opinion, the land reform was aimed at 'lifting the poor up and taking the rich down so that the two classes became equal to each other'.[90]

Although local authorities carried out land redistribution by 'mobilising' (*vận động*) the land-rich households to share some of their land with land-poor and landless households, they faced strong resistance from land-rich peasants and even from some of the intended beneficiaries. A former vice-chairman of Long Điền B commune recalled:

> Land-rich people were dissatisfied with the policy. Even now they still curse us; a few carried long swords to the field to resist sharing their land. But because people at that time feared the new authorities, they did not dare fight us violently. Meanwhile, [poor] peasants were so heavily influenced by capitalist and feudalist thoughts that they refused to receive redistributed land. People said it was weird to take others' land. It was a difficult time for us. Some cadres did not want to share their own land but we did not dare discipline them because our staff members were few. We also did not dare touch land of higher officials for fear of their revenge [*sợ bị trù dập*].[91]

To overcome peasants' resistance, Long Điền B commune authorities decided to carry out land redistribution in a way that tackled the 'easiest first, hardest later'. They focused on redistributing communal and religious land first and then mobilised individuals who had more land (such as more than 10 hectares). At the same time, commune authorities implemented other policies such as adopting high-yielding rice, increasing the number of crops per year (*chuyển vụ*), building

89   Ibid, p.173.
90   Author's interview, 29 July 2005, Long Điền B.
91   Ibid.

production solidarity teams and monopolising the supply of fertilisers and fuel to peasants. These policies also facilitated land sharing among peasants.[92] A former commune official recalled:

> Due to a shortage of fuel for using water pumps, people had to water their fields by scooping. It was impossible for those who had more than 50–100 *công* [5–10 hectares] of land to manage all their land. Therefore, we mobilised those who had more than 50 *công* of land to share their land with others. If they were able to manage all their land, they would not need to share it. But if not, the land would be shared with the land-poor and landless households. No land was allowed to be uncultivated.[93]

As mentioned, however, land redistribution was not well received even by some land-poor peasants. Some beneficiaries refused to accept redistributed land because they felt 'weird' (*kỳ cục*) about taking others' land or were afraid of hurting others' feelings. Some did take the redistributed land but did not dare accept a large amount because they were afraid of being unable to grow high-yielding rice. Some others said they took redistributed land because they did not want to be moved to the new economic zones. An elderly man in Long Quới II hamlet, Long Điền B commune, described his discontent:

> The policy of sharing 'one's rice and clothes' was not suitable for people here. A majority of peasants did not want it. Those who had more land did not want to share some of their land because they had accumulated it with sweat and tears [*bằng mồi hôi nước mắt*]. Those who did not have land did not want to take the land of others. They feared that, when receiving land, they would have to adopt new rice seeds and two rice crops per year with which they were unfamiliar and would be unable to make it profitable.[94]

Meanwhile, some land-rich peasants managed to avoid land redistribution by dividing their fields among their children and relatives. Local officials also encouraged them to do this.[95]

---

92   Ibid.
93   Ibid.
94   Author's interview, 16 June 2005, Long Điền B.
95   Author's interviews, June–August 2005, Long Điền B. Local peasants said that, after reunification, they were informed that the state would take the landless and land-poor to new economic zones. Therefore, to avoid going to these zones, they had to accept redistributed land.

A former production unit cadre in Long Điền B remembered that land redistribution was primarily implemented on the basis that 'the land-rich lent some of their fields [*cho mượn đất*] to land-poor households to make a living'.[96] Other people confirmed that, during land redistribution, land-rich peasants 'lent' them a few *công* of land. When lending the fields, the land-rich peasants often said:

> Now I lend you the area for high-yielding rice cultivation. But if you fail to grow or the state gives up the requirement of growing high-yielding rice and returns to traditional rice [*lúa mùa*], then please return the field to me.[97]

This later had unintended consequences when land disputes developed between new and previous landowners when collective farming was dismantled (discussed in Chapter 6).

Despite the land redistribution in Long Điền B from 1975 to 1978 reducing land inequality among peasants, many landless peasants did not receive any land at all. A landless resident in the commune recalled that he did not receive any land until 1982; he argued that the 'share one's rice and clothes' policy benefited only a small proportion of landless households because most of the land-rich had distributed their land to their children and relatives before the state could touch it.[98] In other words, the main beneficiaries of land reallocations in Long Điền B from 1975 to 1978 were relatives of the land-rich households rather than the land-poor and landless households targeted by VCP policy.

According to a Chợ Mới party cell report, by the end of 1978, district authorities had redistributed about 3,026 hectares of land (about 10 per cent of the cultivated area) to 5,474 landless and land-poor households.[99] In general, the results of land redistribution from 1975 to 1978 in An Giang province were modest. By the end of 1978, the province had taken 20,000 hectares of land from 'landlords and feudalists' to redistribute among the land-poor and landless households.[100]

96   Author's interview, 24 June 2005, Long Điền B.
97   Author's interviews, June–August 2005, Long Bien B.
98   Author's interview, 27 June 2005, Long Điền B.
99   BCHDBHCM, *The History of Chợ Mới Party Cell*, p. 173.
100   STTVHAG, *General Information*, p. 9.

According to a report from the Committee for Southern Agricultural Transformation (BCTNNMN), Vietnam's Southern Region, which included the Mekong Delta and the South-East Region, had confiscated and redistributed 191,931 hectares of 'exploited' land to landless and land-poor households between 1976 and 1978. Of that amount, 'Tiền Giang province had confiscated 12,000 hectares of land from 174 landlords, 468 rich peasants, rural capitalists, and reactionaries'; Long An province confiscated 15,543 hectares of land, Bến Tre province confiscated 55,600 hectares, Đồng Tháp took 13,321 hectares, An Giang 28,800 hectares and Minh Hải 19,814 hectares.[101] The report said:

> Land reform during 1976–1978 focused largely on nationalising the land of foreign farms and confiscating land of landlords, capitalist-compradors and reactionaries … A large proportion of this land was abandoned and occupied illegally.[102]

Moreover, the report revealed that, in many locations, local authorities did not know how to use the confiscated land. Some used it to establish state farms, collectives or production units or lent it to military, state or mass organisations to produce food. Only small amounts of land were used for redistribution among land-poor and landless households, despite the fact these still made up a large proportion of the rural population.[103]

According to a survey carried out by the BCTNNMN, landless and land-poor households still accounted for 18–31 per cent of the rural population and occupied just 10 per cent of the land in July 1978 (see Table 3.2).

In response to this situation, in November 1978, VCP leaders issued Directive No. 57/CT-TW, which called for the continued elimination of the exploitative practices of rich peasants, rural capitalists and vestiges of feudal exploitation. The directive reported:

> In many areas of the south, eliminating the vestiges of feudal and landlord exploitation has not been carried out fully. In many areas [local authorities] do not understand clearly the need to eliminate

---

101 BCTNNMN, *Report on Land Redistribution*, p. 9. It is worth noting that this report gave a higher figure for land redistribution in An Giang (28,800 hectares) than *An Giang*'s report of 20,000 hectares.
102 Ibid.
103 Ibid., pp. 8–9.

rich peasants', rural capitalists' and some upper–middle peasants' exploitation … in many areas, party members who have come from the exploiting class still hold key leadership positions in the commune and hamlet authorities; they have not been enlightened [*giác ngộ*] about the party, nor yet understand clearly the policy of the party-state; even some try to protect the interests of the exploiting class.[104]

It therefore urged local authorities to 'continue to be resolute in eliminating the exploitation of the exploiting class and to share some of their fields with others'. These households were allowed to retain a limited amount of land, equal to the average land per capita in the commune.[105]

Under Directive No. 57, the speed of land redistribution and collectivisation increased in many provinces of the Mekong Delta. However, land redistribution had not been carried out seriously, and local authorities did not follow official regulations nor were they under regular and close leadership.[106]

Table 3.2 Social structure in seven typical hamlets in seven provinces of the Mekong Delta in July 1978

Type of household[1]		An Giang	Đồng Tháp	Long An	Kiên Giang	Minh Hải	Tiền Giang	Bến Tre
I: Non-farming households	Percentage of households	3.32	1.42	1.43	2.73	1.42	8.14	2.67
	Percentage of land	0.35	0.11	0.04	0.10	0.16	0.45	0.45
II: Poor households	Percentage of households	25.50	31.31	20.43	24.20	21.09	18.30	19.30
	Percentage of land	7.70	11.80	5.46	9.21	8.17	6.84	11.90
III: Middle households	Percentage of households	47.00	45.54	56.46	59.79	50.47	63.70	71.88
	Percentage of land	33.00	48.39	47.20	65.13	53.00	67.79	73.00
IV: Upper–middle households	Percentage of households	17.10	16.64	15.23	11.79	20.06	13.00	5.74
	Percentage of land	29.48	28.20	27.30	20.29	33.70	20.40	14.10

---

104 ĐCSVN, Directive No. 57, p. 468.
105 Ibid, pp. 469, 472.
106 BCTNNMN, *Report on Land Redistribution*.

Type of household[1]		An Giang	Đồng Tháp	Long An	Kiên Giang	Minh Hải	Tiền Giang	Bến Tre
V: Rich households	Percentage of households	7.10	4.09	6.45	1.47	2.96	1.86	0.41
	Percentage of land	29.70	11.50	20.00	5.21	4.98	4.60	0.55

[1] Type I: non-farming households; Type II: poor households, including land-poor and landless households who were engaged in wage labour; Type III: middle households who had enough land for their farming needs; Type IV: upper–middle households who had sufficient land, and some of which hired wage labour; Type V: rich households who had much land and many machines and engaged in capitalist business.

Source: Trần Hữu Đính (1994), *Quá trình biến đổi về chế độ sở hữu và cơ cấu giai cấp nông thôn Đồng Bằng Sông Cửu Long (1969–1975)* [*The Process of Ownership and Class Structure Change in Rural Mekong Delta, 1969–1975*], Hà Nội: NXB Khoa Học Xã Hội, p. 103.

Table 3.3 Landholdings in and social composition of a typical hamlet in An Giang in 1978

Household type	Number of households	Percentage of households (%)	Total area of land holdings (ha)	Percentage of holdings (%)	Per capita land (sq m)
I: Non-farming households	15	3.3	3.00	0.31	657
II: Poor households	115	25.5	73.24	7.70	1,151
III: Middle households	212	40.0	314.20	33.04	2,396
IV: Upper–middle households	77	17.1	280.00	29.48	2,774
V: Rich households	32	7.1	280.00	29.47	14,563

Source: Nguyễn Thành Nam (2000), Việc giải quyết vấn đề ruộng đất trong quá trình đi lên sản xuất lớn ở Đồng bằng Sông Cửu Long 1975–1993 [Resolving land issues in the process of large-scale production in the Mekong Delta, 1975–1993], PhD thesis, Đại Học Khoa Học Xã Hội and Nhân Văn, Hồ Chí Minh, p. 47.

Therefore, during the period 1979–81, the 13 provinces in the Southern Region redistributed only 71,292 hectares of land, which accounted for only one-third of land redistribution during the period 1976–78. From 1979 to 1981, An Giang redistributed 6,000 hectares of land (additional to the 20,000 hectares previously redistributed), Hậu Giang redistributed 13,588 hectares and Kiên Giang 4,890 hectares.[107]

107 Ibid, p. 11.

In summary, the land reforms in An Giang and other provinces in the Mekong Delta from 1975 to 1980 did not meet the target of eliminating exploitation and redistributing land to land-poor households. Indeed, land redistribution in An Giang from 1975 to 1980 redistributed only 26,000 hectares of land, which was just 43 per cent of the target (60,225 hectares).[108] Despite this, land reform had significantly weakened the landed class, undermined their capacity to produce commodity rice and transformed the existing land tenure system.

Adopting high-yielding rice and two rice crops per year (*chuyển vụ*)

VCP leaders viewed the Mekong Delta as the 'rice granary' (*vựa lúa*) of the country and considered the region undercultivated and underexploited. After reunification, central authorities sent a group of researchers to the region and their study showed it had great agricultural potential because 'floating rice' and the traditional single rice crop per year were still the main crops there. Two rice crops per year would use only about 250,000 hectares of the delta's 2 million hectares of agricultural land. Therefore, the central government urged farmers in the Mekong Delta to adopt two rice crops per year as early as the winter–spring of 1975–76.[109]

From 1976 to 1980, provincial authorities in An Giang pushed the adoption of high-yielding rice to replace the traditional 'floating rice'. They considered crop transformation essential to facilitate agricultural development, land reform and collectivisation.[110] However, the adoption of high-yielding rice and two crops per year in An Giang province and elsewhere in the Mekong Delta encountered strong peasant resistance. Peasants in the Mekong Delta had a long history of cultivating floating rice (*lúa nổi*), which had adapted well to the annual flooding and other

---

108 In mid-1980, An Giang's leaders announced that the province had essentially completed its land redistribution—of 60,225 hectares (see An Giang hoàn thành cơ bản công tác cải tạo nông nghiệp [An Giang has completed agricultural transformation], *An Giang*, 22 November 1985, p. 1).

109 Phan Quang (1981), *Đồng Bằng Sông Cửu Long* [*The Mekong Delta*], Hà Nội: NXB Văn Hóa, pp. 77–8.

110 According to Võ Tòng Xuân, the area of land used for floating rice in An Giang before 1975 was about 180,000 hectares, accounting for the largest proportion of rice-growing land. Võ Tòng Xuân and Chu Hữu Quý (1994), *Đề Tài KX 08-11: Tổng kết khoa học phát triển tổng hợp kinh tế xã hội nông thôn qua 7 năm xây dựng và phát triển An Giang* [*KX Account 08-11: Summing Up An Giang's Socioeconomic Construction and Development over the Past 7 Years*], Long Xuyên: Chương Trình Phát Triển Nông Thôn An Giang, p. 31.

local ecological and cultural conditions. The productivity of floating rice was lower than the high-yield variety, but more stable—around 10–15 *giạ* (200–300 kg) of rice paddy per *công*. Moreover, floating rice cultivation did not demand a large investment in fertilisers, pesticides, labour and land preparation. With floating rice, a peasant could cultivate a large tract of land with little effort. Local peasants referred to floating rice cultivation as 'caring little but getting a real harvest' (*làm chơi ăn thật*). In March (of the lunar calendar), they sowed rice and then went home until it was time to harvest. They cultivated one rice crop a year and enjoyed a lot of spare time during which they could fish and conduct other economic and cultural activities.[111]

Some peasants in Long Điền B remembered trying before 1975 to adopt high-yielding rice on parts of their land; however, they had little understanding of it. The high-yielding rice required inputs such as fertiliser and pesticides, intensive maintenance and levelling of the land, but they were unable to afford such extras. Even land-rich peasants feared not being able to manage all their fields if they were planted with high-yielding rice. A secondary schoolteacher in Long Điền B who adopted high-yielding rice in 1972 recalled:

> In the past, my family had adopted high-yielding rice no. 8 [*lúa thần nông 8*] on a trial basis. At that time, the land had not been levelled [*chưa bằng phẳng*] and I was busy teaching and did not have much time to care [for it] so I grew only a small amount. Because high-yielding rice had not been adopted extensively in the field, mice and all insects attacked my crops. So, the rice productivity was poor. The highest productivity I gained was about 20 *giạ* per *công* [equal to 4 tonnes per hectare], but it cost me a lot [*chi phí quá nhiều*].[112]

Villagers in Long Điền B commented that, in 1977, soon after people had harvested their traditional rice crop, the local authorities announced the adoption of high-yielding rice and two rice crops per year. Many people refused and continued to cultivate subsidiary crops (*vụ màu*)—mostly watermelon—instead of a second rice crop. In response to peasants' resistance, local authorities set fire to a field to clear the land for crop conversion. An elderly man whose 3 *công* of watermelon was burnt at that time recalled:

---

111 Author's interviews, June–October 2005, Long Điền B.
112 Author's interview, 5 August 2005, Long Điền B.

The authorities did things forcibly [*làm mạnh*]. My watermelon field was growing well and nearly ready for harvest. The morning of that day, my son and I went to water the field as usual. Then in the afternoon the authorities ['*mấy ổng*'] suddenly set the field on fire without informing me. The whole field [*cả cánh đồng*] burned. Many people who lost their watermelon crops shouted and cried [*la chửi và khóc*]. Some had lost 7 to 10 *công* of watermelon. The authorities did an odd thing [*làm kỳ cục*]. They said that they did so in order to plough the field for second rice crop transformation. However, it took three months from firing to ploughing the field. Therefore, people grew more upset.[113]

A former tractor driver who was in charge of ploughing the field for the second rice crop recollected:

At that time, I was a tractor team member [*đội máy kéo*]. Nobody dared to plough the field but I did. Some of my relatives criticised me and considered me a person without ancestors [*người không có ông bà*]. But I knew that we could not refuse to comply with the policy [*chủ trương*]. Adopting two rice crops per year and planting high-yielding rice were compulsory so we had to follow. When I was ploughing the fields, there were some guys who carried long swords [*dao mác*] to block the tractor's path. Frankly speaking, I did not dare plough the field without the support of authorities. At that time, officials from the commune's agricultural department, commune police and even the commune chairman himself came to support us. Without them, nobody dared plough … those who disagreed with the policy tried to intimidate us rather than openly confront us because they too were afraid of the authorities. Everyone was afraid to upset the Vietnamese communist cadres [*nói đến ba ông Việt Cộng ai cũng sợ*].[114]

Despite strong peasant resistance, the authorities used various measures to force the adoption of high-yielding rice, such as prohibiting peasants from farming outside their residential area, land redistribution and controlling the supply of rural goods and inputs. As a result, the adoption of two crops per year in An Giang increased the area under rice cultivation from 31,509 hectares in 1976 to 79,066 hectares in 1980.[115] In Chợ Mới district, it increased from 3,120 hectares in the winter–spring of 1976–77 to 16,430 hectares in the winter–spring of 1978–79, which accounted for nearly half of the total rice land in the district. Some communes in Chợ Mới district, such as Hòa Bình, Nhơn

---

113 Author's interview, 29 June 2005, Long Điền B.
114 Author's interview, 30 June 2005, Long Điền B.
115 Võ Tòng Xuân and Chu Hữu Quý, *KX Account 08-11*, p. 31.

Mỹ and Hội An, had by this time completed their adoption of high-yielding rice and the two-crop requirement.[116] Phú Tân district, in which the local authorities were weak and 90 per cent of the population was Hòa Hảo, implemented crop transformation extensively, too. The area of high-yielding rice there increased from 6,600 hectares in 1975 to 17,500 hectares in 1980.[117]

Conclusion

After the war, in response to the VCP's agrarian policies, local authorities in QN-ĐN in the Central Coast and An Giang in the Mekong Delta focused on resolving postwar problems and preparing for collectivisation. Despite Thăng Bình district being heavily damaged by the war, the new authorities there and in many other districts of QN-ĐN swiftly consolidated their power and were able to implement the main contents of the VCP's post-1975 reforms. The new authorities in Chợ Mới and other districts of An Giang still faced difficulties in building government and implementing the VCP's policies.

There are at least two main reasons for the 'better' results in implementation in QN-ĐN than in An Giang. First, after the war, QN-ĐN had a larger number of ex-revolutionaries and southerners returned from the north than did An Giang. QN-ĐN cadres at the district, commune and village level had more experience with the VCP's policies and northern collectivisation and were more loyal to the party's socialist transformation of agriculture than their counterparts in An Giang. For instance, authorities in QN-ĐN carried out preparatory measures for collectivisation forcefully and simultaneously. Meanwhile, An Giang authorities implemented these policies more cautiously. Peter Nolan also shows that the relative strength and quality of the Communist Party apparatus at the village level is one reason for more socioeconomically successful collectivisation in China than in the former Soviet Union.[118]

---

116 BCHDBHCM, *The History of Chợ Mới Party Cell*, p. 171; STTVHAG, *General Information*, p. 14.
117 Phú Tân đẩy mạnh phong trào hợp tác hóa nông nghiệp [Phú Tân intensifies collectivisation], *An Giang*, 27 November 1980, p. 1.
118 Nolan, Collectivization in China, pp. 195–7.

Second, the consequences of war in QN-ĐN were more severe than in An Giang. After the war, most peasant households in QN-ĐN were extremely poor and the social and economic structure of rural communities was flattened and relatively homogeneous. Most peasants were engaged in subsistence production and struggled to make a living. Given the extremely difficult conditions in QN-ĐN, most poor and powerless villagers tended to comply with the new agrarian policies to avoid any political, social or economic disadvantages imposed by those in power. In addition, some of the new policies, such as labour exchange teams and land sharing, seemed to fit well with local culture and practices. Because of the absence of market relations, cultural patterns of behaviour such as reciprocity and labour exchange were popular in QN-ĐN.

Meanwhile, because the consequences of war in Chợ Mới, in An Giang, were less severe than those in QN-ĐN, peasant households were better off and lived in more open, highly stratified and occupation-diverse rural communities. An Giang villagers therefore had greater capacity to evade and resist state policies that were unattractive to them. Moreover, some of the new agrarian policies—such as production solidarity teams, land redistribution and prohibition of non-resident cultivators—did not fit with local practices and conditions in which market relations and private land tenure were well established. These policies therefore encountered strong peasant resistance in An Giang and elsewhere in the Southern Region (discussed further in the next chapter).

4

Establishing collective organisations, 1978–81

Introduction

After the war, along with restoring the country's war-torn economy, Vietnamese Communist Party (VCP) leaders put great effort into preparing the south for collectivisation. They urged local authorities to carry out collectivisation step by step, moving from a low to a high level and from simple to complicated forms of collective organisations. The party urged local authorities to experiment with pilot collectives before expanding collectives extensively in the south. In 1977, VCP leaders instructed each province in the south to select one district in which to build a pilot collective in the winter–spring season of 1977–78. They were determined to carry out collectivisation and planned to complete it by 1980.[1]

The pilot collectives failed, especially in the Southern Region, and their form and character had to be changed to fit regional conditions. Despite these failures, party leaders praised the success of the pilot collectives in terms of mobilising peasants, collectivising their means of production and increasing productivity and collective members'

---

1   Ban Chấp Hành Trung Ương [hereinafter BCHTU] (1977), *Chỉ thị 29-CT/TW về chính sách được áp dụng ở các hợp tác xã thí điểm ở Miền Nam* [Directive No. 29-CT/TW on Policy for Pilot Collectives in the South], 26 December, Hà Nội: NXB Nông Nghiệp.

incomes compared with individual farming.[2] After the conference to consolidate agricultural collectives in the south in August 1978, VCP leaders persistently pushed collectivisation to achieve its goals.[3] Moreover, the border war with Cambodia and China helped increase the VCP's determination to accomplish collectivisation.

Results varied, however, from region to region, and major problems were encountered, especially in the Mekong Delta. By the end of 1979, collectivisation was essentially complete in the Central Coast region, while it had been achieved for only a modest proportion of land and peasant households in the Mekong Delta.

This chapter examines the process and performance of pilot collectives in the Central Coast and the Mekong Delta and how and why VCP leaders decided to accelerate collectivisation; why collectivisation occurred rapidly in the Central Coast but faced many difficulties in the Mekong Delta; and how local authorities, cadres and peasants in both regions reacted to and influenced the collectivisation process.

Experiments with collectives in Quảng Nam-Đà Nẵng (QN-ĐN), Central Coast region

Two years after resolving their postwar problems, QN-ĐN's provincial leaders seemed to trust their ability to meet not only the food subsistence requirements of the province, but also to bring the province's agriculture towards socialist large-scale production.[4] In September 1977, the QN-ĐN party cell released its resolution 'on development and agricultural transformation' (Nghị quyết về phát triển và cải tạo nông nghiệp), which scheduled the building of pilot collectives in 1977, their extension in 1978 and accelerating and completing collectivisation by 1980.[5] Provincial leaders argued that establishing collectives in rural areas

---

2   Nguyễn Thành Thơ (1978), Ra sức tiến hành hợp tác hóa nông nghiệp [Do our best to implement collectivisation], in Võ Chí Công et al. (eds), *Con đường làm ăn tập thể của nông dân* [*The Collective Farmer's Way*], Hồ Chí Minh: NXB Tp. Hồ Chí Minh, p. 13.
3   Ibid., p. 13. According to Nguyễn Thành Thơ, by August 1978, 132 pilot collectives had been established in the south. However, the majority of these (108) were in the Central Coast region; the Mekong Delta had only two, the South-Eastern region had 12 and the Central Highlands had 18.
4   This statement can be found in *Quảng Nam-Đà Nẵng* (Be victorious in agriculture, 26 April 1977, p. 1).
5   Nghị Quyết hội nghị Ban chấp hành đảng bộ tỉnh (khóa 11) về vấn đề phát triển và cải tạo nông nghiệp [Resolution of the Eleventh Provincial Party Congress on agricultural transformation and improvement], *Quảng Nam-Đà Nẵng*, 7 September 1977, p. 1.

would not only fulfil the aims and ideals of the Communist Party, but would also be the peasants' reward for enduring great losses during the country's wars.[6]

As an initial step, provincial leaders decided to build three pilot collectives in three different districts: a commune-sized pilot collective (Duy Phước collective) in Duy Phước commune, Duy Xuyên district; a half-commune–sized collective (Hòa Tiến 1 collective) in Hòa Tiến commune, Hòa Vang district; and a commune-sized pilot collective (Bình Lãnh collective) in Bình Lãnh commune, Thăng Bình district. While the first two collectives were in lowland areas where land was more fertile and peasants were more prosperous, the third was in an area of undulating, less fertile land.

The main criteria provincial leaders used were that the place had to be an ex-revolutionary base (cơ sở cách mạng) and its party cell had to be 'strong' and decisive. For example, Bình Lãnh commune was selected because the area had been a strong revolutionary base and its party cell was loyal to the party's agricultural transformation policy, which would enable it to build a successful pilot collective.[7]

To establish these pilot collectives successfully at the outset, QN-ĐN authorities undertook considerable preparations. They set up a provincial committee responsible for agricultural transformation (Ban cải tạo nông nghiệp)[8] and launched a series of campaigns urging local cadres and peasants to study the provincial party committee's resolution.[9] In addition, the authorities opened a collectivisation school (trường hợp tác hoá) to train collective cadres. It trained 37 cadres and eight accountants for Duy Phước collective, 17 cadres and six accountants for Bình Lãnh collective and 15 cadres and five accountants for Hòa Tiến collective.[10] A former vice-chairman of Bình

---

6   This was cited from the speech by Hồ Nghinh, the Communist Party secretary for the province, at the meetings to establish Hòa Tiến 1 pilot collective and Bình Lãnh pilot collective (Mở đại hội xã viên thành lập hợp tác xã nông nghiệp Bình Lãnh và Hòa Tiến [Members' congress held to establish Bình Lãnh and Hòa Tiến collectives], *Quảng Nam-Đà Nẵng*, 5 November 1977, p. 1).

7   Author's interview, 20 October 2004, Bình Lãnh commune.

8   Thành lập Ban cải tạo nông nghiệp [Establishing a committee for agricultural transformation], *Quảng Nam-Đà Nẵng*, 4 October 1977, p. 1.

9   Phấn khởi nghiên cứu học tập Nghị quyết Tỉnh ủy về phát triển và cải tạo nông nghiệp [Studying the Provincial Party Committee's resolution on agricultural transformation], *Quảng Nam-Đà Nẵng*, 10 September 1977, p. 1.

10   Tổng kết xây dựng thí điểm hợp tác xã nông nghiệp [Summing up establishing pilot collectives], *Quảng Nam-Đà Nẵng*, 27 May 1978, p. 1.

Lãnh collective remembered that, before establishing pilot collectives, he and other cadres in Bình Lãnh were sent for four months of training in Đà Nẵng city.[11]

Furthermore, to prevent peasants from slaughtering and selling draught animals before the collectives were established, the provincial people's committee released a directive forbidding people from buying draught animals. The directive stipulated that

> buying draught animals within the commune requires the permission of the local People's Committee [Uỷ ban Nhân dân xã]; exchange of stocks between two communes requires permission from the district People's Committee; and buying and selling animals between two districts requires permission from the provincial People's Committee.[12]

It also forbade peasants from intentionally injuring, poisoning or slaughtering their draught animals. Each commune also established a committee for mobilising peasants into collectives (Ban vận động thành lập hợp tác xã), which were in charge of convincing peasants to join collectives through study meetings or visiting 'difficult' peasant households who were reluctant or refused to join collectives.[13]

Villagers in Bình Lãnh collective recalled that, to establish the pilot collectives there, central, provincial and district authorities provided considerable assistance and resources to the commune. Seven northern cadres, including the chairs of the 'advanced' collectives in Thanh Hóa province, came to stay in the commune for seven months to provide help. They even directly managed the collectives. The district's authorities sent a vice-chairperson of its economic department to work as a chairperson of the Bình Lãnh collective. The central and provincial governments had also invested a great deal in Bình Lãnh collective by providing hundreds of tonnes of cement, lime, fertiliser and other resources.[14]

After a few months of preparation, provincial authorities started to establish the pilot collectives. An October 1977 article in the Quảng Nam-Đà Nẵng newspaper reported that 100 per cent of peasant

---

11   Author's interview, 14 October 2005, Bình Lãnh.
12   Prohibiting private merchants from purchasing and slaughtering livestock, Quảng Nam-Đà Nẵng, 24 September 1977, p. 1.
13   Author's interviews, October–December 2005, Bình Lãnh and Bình Định communes.
14   Author's interviews, October–December 2005, Bình Lãnh.

households in Duy Phước commune, 96 per cent in Bình Lãnh and 95 per cent in Hòa Tiến 1 had 'voluntarily signed the form to participate in collectives'. The article urged 'peasants in the province to follow the path of collective farming of Duy Phước, Bình Lãnh and Hòa Tiến'.[15] In late October 1977, local authorities announced the completion of the three pilot collectives and the establishment of collective members' congresses (*đại hội xã viên*) to select their managerial boards (*ban quản trị hợp tác xã*). The percentage of peasant households joining collectives increased—to 98.3 per cent in Bình Lãnh collective and 99 per cent in Hòa Tiến 1.[16] Almost all land, draught animals and other means of production were collectivised: 'One hundred per cent of agricultural land in Duy Phước [704 hectares], 97.7 per cent in Bình Lãnh [562 hectares] and 100 per cent in Hòa Tiến [373 hectares].' Approximately 87 per cent of draught animals in Duy Phước, 90.1 per cent in Bình Lãnh and 100 per cent in Hòa Tiến 1 were collectivised, as were 'production tools, machinery and other implements necessary for collectives'.[17]

Although the official documents claimed that most peasants joined pilot collectives voluntarily, few residents recalled being enthusiastic participants. Many peasants in Bình Lãnh claimed they did not like joining the collective but had to do so. They were 'coerced' (*bị bắt buộc*) or 'pressured' (*bị bắt bí*) and fearful of being isolated (*sợ cô lập*) if they did not join. Some were indifferent and just followed what others did. Some—especially land-poor but labour-rich families—seemed to be more eager than others to join. Some decided to join because they believed the state would take care of them and not let them die of hunger regardless of the collectives' performance.[18] The provincial newspaper reported that some peasants did not believe in collective farming and, in the first few months, the pilot collectives found it difficult to mobilise members to work in the fields. Moreover, some peasants engaged in obstructive practices such as selling their draught animals before collectivisation, renting land in uncollectivised communes or seeking

---

15   Bà con nông dân trong tỉnh hãy theo con đường làm ăn tập thể của Duy Phước, Bình Lãnh, Hòa Tiến [Peasants in the province should follow the collective farming paths of Duy Phước, Bình Lãnh and Hòa Tiến people], *Quảng Nam-Đà Nẵng*, 11 October 1977, p. 1.

16   Mở đại hội xã viên thành lập hợp tác xã nông nghiệp Duy Phước [Members' congress held to establish Duy Phước collective], *Quảng Nam-Đà Nẵng*, 29 October 1977, p. 1; Members' congress held to establish Bình Lãnh and Hòa Tiến collectives, *Quảng Nam-Đà Nẵng*, 5 November 1977, p. 1.

17   Summing up establishing pilot collectives, *Quảng Nam-Đà Nẵng*, 27 May 1978, p. 1.

18   A woman in Bình Lãnh argued: 'entering [a] collective means hunger, but the party could not let people die of it' (Author's interview, 20 October 2005).

outside jobs to make a living.[19] While the collective transplanted seedlings, many people focused only on growing their subsidiary crops (*màu*) on their '5 per cent land' (the portion of their land they were allowed to retain) or on land the collective had not yet used.[20]

Despite many peasants not liking collective work, the establishment of the three pilot collectives in QN-ĐN did not face difficulties. Local authorities were able to take control of peasants' land and other means of production and mobilise peasants into collective work. As in QN-ĐN, other parts of the Central Coast confronted few difficulties in building their pilot collectives. For example, Nghĩa Bình province, neighbouring QN-ĐN, also succeeded in building pilot collectives, among them one in Nghĩa Lâm commune that 99.9 per cent of peasant households joined.[21]

In May 1978, when the pilot collectives had their first harvest, the *Quảng Nam-Đà Nẵng* newspaper reported 'a victory in the first step' (*thắng lợi bước đầu*). Staple food production and productivity in the three pilot collectives, it said, reached the highest figure ever, and the income of the collectives' members was higher than that of individual farmers.[22]

At a conference to provide a summary of the state of the pilot collectives in May 1978, Hồ Nghinh, the Communist Party Secretary for QN-ĐN, praised the collectives' achievements:

> [T]hat victory confirmed the correct ... policies of our party, reflected the superiority of the mode of socialist collective production ... that victory defeated the propaganda and distorted statement of the enemy as well as solved doubts [*hồ nghi*] and anxiety [*băn khoăn*] of some cadres and peasants.[23]

---

19 Summing up establishing pilot collectives, *Quảng Nam-Đà Nẵng*, 27 May 27, p. 1.

20   Thắng lợi bước đầu của phong trào Hợp tác hóa nông nghiệp [The first victory steps of collectivisation], *Quảng Nam-Đà Nẵng*, 13 May 1978, p. 1.

21   HTX Nghĩa Lâm (1978), Kinh nghiệm xây dựng hợp tác xã Nghĩa Lâm, tỉnh Nghĩa Bình [Experiences from establishing Nghia Lam collective in Nghĩa Bình province], in Võ Chí Công et al. (eds), *Con đường làm ăn tập thể của nông dân* [*The Collective Farmer's Way*], Hồ Chí Minh: NXB Tp. Hồ Chí Minh, p. 147. Nghĩa Bình was the result of the amalgamation of Quảng Ngãi and Bình Định provinces.

22   The first victory steps of collectivisation, *Quảng Nam-Đà Nẵng*, 13 May 1978, p. 1.

23   Hồ Nghinh: Thắng lợi của việc xây dựng thí điểm hợp tác là thắng lợi có ý nghĩa của toàn đảng bộ và nhân dân toàn tỉnh [Hồ Nghinh: The success of pilot collectives is a significant victory for the province's party and people], *Quảng Nam-Đà Nẵng*, 27 May 1978, p. 1.

Peasants in Bình Lãnh collective confirmed that, in the first season, the collective had a bumper harvest. Collective members received 3 kilograms of paddy per workday—a level never repeated in later years of collective farming. Many villagers attributed the bumper crop to favourable weather, the huge investment in the pilots and 'good soil' because the land had been uncultivated for a long time.[24] Some peasants and former staff of collectives revealed that the high payment per workday the peasants received for the first season was an inflated figure that the authorities used to attract peasants in other places to join collectives. To increase the payment per workday for peasants, leaders of the pilot collectives had transferred some of the peasants' current work-points to the next season. A former cadre of Bình Lãnh collective confirmed this deception:

> The payment per workday was actually about 2 kilograms at that time. However, by trickily transferring part of the amount of work-points [ghế diểm] to the next season, payment per workday reached 3 kilograms. That's why we never achieved that figure.[25]

In short, although peasants in QN-ĐN and other parts of the Central Coast were not eager to join collectives, the pilot collectives there faced weaker resistance than in other regions in the south. Local authorities were able to collectivise peasants' land and their other main means of production and mobilise them to undertake collective work.

Building pilot collectives in the Mekong Delta

Communist Party leaders anticipated strong peasant resistance to collectivisation in the Southern Region, so they undertook cautious experiments with pilot collectives. Instead of requesting each province to build its own pilot collectives, as in the Central Coast, in the Mekong Delta, party leaders built only one pilot collective for the whole region—in Tân Hội commune, Cai Lậy district, Tiền Giang province—in February 1977. Tân Hội commune was selected because it had been a revolutionary base of the National Liberation Front (NLF).

---

24 Villagers said the soil was in good condition because it had been left uncultivated for a long time. The collective also used huge volumes of agricultural inputs such as fertiliser, lime and manure.
25 Author's interview, 20 October 2005, Bình Lãnh.

This pilot was a commune-sized collective called Tân Hội collective (*Hợp Tác Xã Tân Hội*), which contained 904 households and 525 hectares of land. To make Tân Hội collective a shining example for the whole region, authorities had to invest considerable resources. For example, the central government sent more than 100 cadres from 'advanced collectives' (*hợp tác xã tiên tiến*) in the north to help. Despite this, the collective faced significant difficulties and many members dropped out. By 1978 only 234 peasant households remained in the collective. To try to save the collective, authorities decided to divide it into two; however, neither collective was able to hold out and both were dismantled (*tan rã*).[26]

With the failure of this large-scale (*qui mô lớn*) collective, party leaders decided to try a smaller-scale pilot. They chose Phú Quới hamlet, Yên Bình commune, in Gò Công district of Tiền Giang province, in which to build a hamlet-sized pilot, called Phú Quới collective. The main criteria for selecting Phú Quới were: 1) the natural conditions were favourable for the adoption of intensive farming (high-yielding rice) and increasing the number of crops per year; 2) peasants in the hamlet had been trained in production solidarity teamwork; and 3) its cadres and mass organisations were strong and capable of building a successful collective.[27]

After one month of preparation, Phú Quới pilot collective was officially established on 17 May 1978. It had 257 households (98.4 per cent of total households), 309.84 hectares of land (97.4 per cent of the total) and nearly 100 per cent of the machinery and draught animals in the hamlet. Unlike the previous experiments, the Phú Quới collective was able to survive beyond a few months of operation and was considered an exemplary example for other provinces.[28] Learning from this experience, some other provinces in the Mekong Delta shifted to experimenting with small-scale pilot collectives and production units (*Tập đoàn sản xuất*).[29]

---

26 Huỳnh Thị Gấm, Socioeconomic changes in the Mekong Delta, p. 80.
27   Đảng bộ huyện Gò Công (1978), Vận động thành lập hợp tác xã thí điểm ở Gò Công [Mobilising and establishing pilot collectives in Gò Công], in Võ Chí Công et al. (eds), *Con đường làm ăn tập thể của nông dân* [*The Collective Farmer's Way*], Hồ Chí Minh: NXB Tp. Hồ Chí Minh, pp. 129–30.
28 Ibid., pp. 133–4.
29 Huỳnh Thị Gấm, Socioeconomic changes in the Mekong Delta, p. 80.

Experimenting with collectives in An Giang

In October 1978, An Giang province began to build pilot collectives. Among them were the Hòa Bình Thạnh commune-sized collective in Châu Thành district, and the Tây Huế hamlet-sized collective in Long Xuyên town. Despite direct assistance from the provincial government, the building of the Hòa Bình Thạnh collective failed in terms of peasants' participation and performance. A recent official document revealed that

> in order to mobilise peasants into joining the collective, policemen had to stand at the edges of the rice fields and request peasants to destroy their vegetable crops in order to give the land to the collective.[30]

Authorities also faced great difficulties convincing peasants to hand over their machinery and tools to collective organisations. The collective had to force peasants to hand over this equipment (*tập thể hoá bằng mọi giá*). Some owners strenuously objected, and 'intentionally removed some machine accessories; some broke the machines' chains or axles before handing them to the collective'. After collectivisation, the machines' new managers lacked the skill and motivation necessary to look after them, so that,

> after one season, 100 per cent of machines were broken and had to be put into storage; hundreds of hectares of land were not ploughed in time and left uncultivated.[31]

As a result, the collective's performance was very poor. Peasants received a low income, equal to just one-quarter of their previous income from individual farming.[32] Thus, Hòa Bình Thạnh collapsed.

Meanwhile, the Tây Huế hamlet-size collective was able to continue despite its poor performance in terms of the quality of peasant work and paddy productivity. When the collective was established, 211 out of 244 households joined, but the collective faced a high dropout rate.[33]

---

30   Xuân Thu and Quang Thiện (2005), Đêm trước đổi mới: Công phá 'lũy tre' [On the eve of the renovation: Breaking through the 'bamboo hedges'], *Tuổi trẻ Online*, 14 December, available from: tuoitre.vn/tin/chinh-tri-xa-hoi/phong-su-ky-su/20051204/dem-truoc-doi-moi-cong-pha-luy-tre/111625.html (accessed 4 October 2017).

31   Ibid.

32   Ibid.

33   Hợp tác xã Tây Huế qua 6 năm làm ăn tập thể [Tây Huế collective over the past 6 years], *An Giang*, 30 December 1983, p. 2.

In general, pilot collectives were not successful in An Giang and other Mekong Delta provinces. Central and provincial leaders in the delta therefore shifted full collectivisation from collectives to production units, which they now considered a basic form of collectivisation.[34]

Experimenting with production units

Realising the failure of pilot collectives, many districts in An Giang built the more modest production units instead. The size of these units ranged from 40 to 50 hectares of land and from 50 to 100 households.

In late 1979, Chợ Mới district chose Long Điền B commune in which to build pilot production units.[35] According to a former official of Long Điền B commune, with the direct assistance of some local and northern cadres from the province of Thanh Hóa, they decided to build pilot production unit no. 1 (*tập đoàn sản xuất số 1*) in Long Phú 1 hamlet. Long Phú 1 hamlet was considered to have strong leadership (*chính quyền ấp mạnh*) thanks to the hamlet chief, who was 'powerful' and 'enthusiastic' (*nhiệt tình*) about agricultural transformation. In addition, a large proportion of the population in Long Phú 1 was Catholic, and most of the land in the hamlet had been owned by the local church (*đất ông cha, đất nhà chung*) but rented to peasants, meaning the land there was already considered more like communal than individual land. Authorities believed that when the church leaders agreed to hand the land over to them, such collectivisation would be acceptable to the peasants. Finally, the proportion of landless and land-poor households in Long Phú 1 was relatively high compared with other hamlets, so authorities expected such households would be more eager to farm collectively than better-off households.[36]

Local villagers recalled that the district committee for agricultural transformation undertook a lot of preparation before establishing production unit no. 1, such as selecting good cadres to fill the unit's management board, rezoning land boundaries and mobilising peasants to join. Peasants in the hamlet whose land was within the boundaries

---

34 Quang Truong, Agricultural collectivization, p. 191.
35 BCHDBHCM, *The History of Chợ Mới Party Cell*, p. 175.
36 Author's interview, 29 July 2005, Long Điền B.

of the production unit were preferred members, while landless households in the hamlet or neighbouring hamlets were also invited to join.[37]

Marshalling peasants to join the production unit was not easy. Villagers recalled that, of the 10 households invited, only five or six participated; more poor than better-off households joined; and 'some better-off households who had more than 30 *công* of land detested [*chê*] collective farming in the production unit, so they ran off to hire land in other places to make a living'. They also recalled that some residents who lost land were so upset they refused to join.[38]

When asked why they joined, many former members of the production unit said it was 'for fear of the new authorities' (*sợ chính quyền mới*), 'fear of being taken to the new economic zones' (*sợ đưa đi vùng kinh tế mới*), because they were being coerced (*bị ép buộc*) or 'in order to keep the land' (*vào để giữ đất*). One man in Long Phú commented:

> [O]ut of 100 people, only five who were landless and loved farming were happy to join. The remainder were coerced into joining; if we didn't participate how could we make a living and keep our land?[39]

Despite such difficulties, 83 households were reported to have joined production unit no. 1, and it began to operate with 55 hectares of land in the summer–autumn of 1979.[40] Unlike a collective, the production unit collectivised land but not machinery, draught animals or other peasant-owned resources. A machinery unit (*tập đoàn máy*) was in charge of the peasants' farm machinery. Villagers referred to the main unit as the 'land unit' (*tập đoàn đất*) to distinguish it from the machinery unit, and both were to become teams or brigades in a future collective. The production unit operated according to a work-points system in which peasants farmed collectively and were rewarded with a number of points. Although official policy encouraged payment for land (*trả hoa lợi ruộng đất*), the leaders of production unit no. 1 did not apply this.

---

37   Ibid. The production unit's management board consisted of five staff who were considered to have come from 'revolutionary tradition–related families' (*gia đình có truyền thống cách mạng*): the chairperson, one vice-chairperson in charge of labour management, another vice-chairperson in charge of planning, one accountant and one storekeeper.
38   Author's interviews, 2–30 June 2005, Long Điền B.
39   Author's interview, 25 September 2004, Long Điền B.
40   The total number of households within the boundaries of the production unit was about 150 (Author's interview, 17 August 2005, Long Điền B).

Despite considerable assistance from district and northern cadres and significant investment in the pilot production unit, its performance was not good. Many people were reported to have 'joined the production unit but did not go to work in the fields' (*vào tập đoàn nhưng không ra đồng*); some sent their children and other 'subsidiary labour' (*lao động phụ*) to work while 'the main labour' (*lao động chính*) in households made a living in other ways. Villagers also mentioned that peasants did their production unit work unenthusiastically and carelessly, and 'no-one took care of common property' (*cha chung không ai khóc*). Moreover, the production unit was unable to mobilise peasants to complete their work on time (*làm không kịp việc*). For example, weeding of the production unit's rice fields went so slowly that 'the weeds grew faster than the speed of weeding'.[41] One man outside production unit no. 1 recalled:

> I went to see how people in the production unit worked collectively. When I saw the weeds were overgrown, I lost my interest [*thấy mà mất ham*] in collective farming. Meanwhile, in individuals' rice fields, I could not see any weeds [*không thấy một cọng cỏ*]. At that time I was afraid that collective farming would expand into my hamlet.[42]

(See Chapter 5 for more about peasants' everyday politics.)

According to a former cadre of production unit no. 1, for the first season (summer–autumn 1979), paddy productivity was about 60 per cent of individual farming rates. Because few peasants were working in the fields, the total of peasants' work-points was small. As a result, payment per workday for peasants was relatively high (about 10 kilograms per workday). Therefore, production unit no. 1 was known as an exemplary case in the Chợ Mới district. Party leaders presented it as a typical case (*đi báo cáo điển hình*) at provincial and central government conferences on pilot collectives in 1979.[43]

In the winter–spring of 1979–80, authorities in Long Điền B commune decided to create another production unit, unit no. 2, in Long Phú 2, at the nearby hamlet of Long Phú 1. Production unit no. 2 faced similar difficulties in mobilising and managing peasants and its performance

---

41 Author's interviews, 27–30 June 2005, Long Điền B.
42 Author's interview, 16 August 2005, Long Điền B.
43 Author's interview, 17 August 2005, Long Điền B.

was poor from the outset. Although the performance of these two pilot collectives steadily deteriorated, local cadres tried their best to keep them from collapsing.

Accelerating collectivisation

In April 1978, VCP leaders released Directive No. 43/CT-TW (14 April 1978), which stressed 'firmly grasping the task of agricultural transformation and speeding it up in the south'.[44] It advocated for local officials in the south to make agricultural transformation their 'central and regular task' (nhiệm vụ trọng tâm và thường xuyên), which they should carry out in a 'positive, unhesitant and not overhasty and careless manner'.[45]

At the conference on the consolidation of agricultural collectives in the south in August 1978, party leaders also revealed that the country now faced a 'new situation' relating to China 'inciting' Cambodia into a border war against Vietnam. The party leaders stressed 'this new situation requires us to speed up agricultural transformation and try our best to implement it in the south'.[46] They planned to implement a great wave of collectivisation in the south in 1979 to complete the establishment of collectives and production units by 1980.[47] The following sections examine how collectivisation was accelerated in QN-ĐN in the Central Coast and in An Giang in the Mekong Delta.

QN-ĐN in the Central Coast

Following the three initial pilot collectives, QN-ĐN established four more (hợp tác xã nông nghiệp), in the spring–summer of 1978. According to Quảng Nam-Đà Nẵng newspaper accounts, by June 1978, the province had established seven pilot collectives, which were reportedly operating well. The provincial leaders attributed the 'good' performance of these pilots largely to 'the correctness of agricultural transformation policy' and 'the superiority of new production

---

44   ĐCSVN, Politburo's Directive No. 43, pp. 183–91.
45   Ibid.
46   Ibid, p. 9.
47   Võ Chí Công (1978), Con đường làm ăn tập thể của nông dân [The collective farmer's way], in Võ Chí Công et al. (eds), Con đường làm ăn tập thể của nông dân [The Collective Farmer's Way], Hồ Chí Minh: NXB Tp. Hồ Chí Minh, p. 71.

relations'.[48] Excited with the performance of pilot collectives and in response to the VCP's Directive No. 43 (14 April 1978), QN-ĐN's leaders called for a rapid and extensive increase in collectivisation for the winter–spring of 1978–79.[49]

By October 1978, QN-ĐN had established 113 collectives involving 46 per cent of the province's peasant households and 35 per cent of its agricultural land.[50] By that time, officials in Duy Xuyên, a key district that had established the first pilot collective in QN-ĐN, announced that collectivisation there was largely complete, making it the first district in QN-ĐN, and the first in the south, to achieve completion.[51] Duy Xuyên had established 19 collectives, which almost all of the peasant households in the district had joined.[52] Inspired by the high speed of collectivisation, provincial leaders decided in October 1978 to shorten by one year the schedule for fulfilling the main targets of their five-year agricultural plan for 1976–80. This meant completing collectivisation and reaching the production target of 500,000 tonnes of staple food by 1979 instead of 1980.[53]

By April 1979, one year after the VCP issued Directive No. 43, QN-ĐN had established 164 collectives, involving 70 per cent of total peasant households. Collectivisation in other Central Coast provinces was also

---

48 Xã Luận: Xây dựng quan hệ sản xuất mới trong nông nghiệp [The editorial: Building new production relations in agriculture], *Quảng Nam-Đà Nẵng*, 14 June 1978, p. 1. The four new pilot collectives were Hòa Tiến no. 2 collective in Hòa Vang district, Quyết Tiến collective and Tiến Phong collective in Điện Bàn district, and Tam Thành collective in Tam Kỳ district.

49 Tích cực chuẩn bị mở rộng phong trào tổ chức hợp tác xã sản xuất nông nghiệp [Be positive towards the extension of collectivisation], *Quảng Nam-Đà Nẵng*, 10 June 1978, p. 1.

50 Thành lập xong 98 hợp tác xã sản xuất nông nghiệp [98 agricultural collectives have been established], *Quảng Nam-Đà Nẵng*, 11 October 1978, p. 3.

51 Đoàn cán bộ Ban cải tạo nông nghiệp Trung ương, các tỉnh Miền Trung và Hội liên hiệp phụ nữ Việt Nam thăm huyện Duy Xuyên [Cadres of the Central Agricultural Transformation Committee, the Central Coast provinces and Vietnam Women's Union visited Duy Xuyên district], *Quảng Nam-Đà Nẵng*, 25 October 1978, p. 1.

52 Duy Xuyên had 28,000 hectares of uncultivated land, 11 communes and 19,462 households. It had 7,000 hectares of agricultural land, including 3,400 hectares of rice-growing land, and the level of land per capita was 1 *sào* and 11 *thước* (equal to 867 sq m). Farming households accounted for 81.4 per cent of the population; fishing, 3.4 per cent; handicrafts, 9.4 per cent; traders, 1.4 per cent; and other professions, 4.1 per cent (Duy Xuyên khẩn trương xây dựng huyện để chỉ đạo và quản lý các hợp tác xã [Duy Xuyên district's rush to build capacity to lead collectives], *Quảng Nam-Đà Nẵng*, 25 October 1978, p. 1).

53 Xã luận: Phấn đấu hoàn thành những mục tiêu về sản xuất và cải tạo nông nghiệp đểa trong năm 1976–1980 của tỉnh vào năm 1979 [The editorial: Do our best to meet 1976–1980 targets of production and agricultural transformation by 1979], *Quảng Nam-Đà Nẵng*, 21 October 1978, p. 1.

rapid. By April 1979, the Central Coast region had largely completed collectivisation in two forms, collectives and production units (see Table 4.1).[54]

Table 4.1 The number of collectives and the percentage of peasant households joining collectives in five Central Coast provinces by April 1979

Province	Number of collectives	Percentage of total peasant households
Bình Trị Thiên	231	85.5
Quảng Nam-Đà Nẵng	164	70.0
Nghĩa Bình	246	57.8
Khánh Hòa	180	80.0
Thuận Hải	183	55.4
Central Coast region	1,004	70.3

Source: Cùng với cả Miền Nam tỉnh ta khẩn trương hoàn thành hợp tác hóa nông nghiệp [Our province, together with southern provinces, hurries to complete collectivisation], *Quảng Nam-Đà Nẵng*, 27 June 1979, p. 1.

Nhân Dân newspaper reported on 29 April 1980 that, by the end of 1979, the Central Coast had established 1,114 collectives and 1,500 production units, which accounted for 83 per cent of peasant households and 76 per cent of agricultural land. The article praised collective farming for achieving better levels of paddy productivity, staple food production and food contribution to the state than individual farming.[55]

By the end of 1979, QN-ĐN authorities announced the completion of collectivisation in the lowland and midland areas. The province had established 235 collectives involving 18,400 peasant households (nearly 93 per cent of the province's peasant households) and 106,000 hectares of agricultural land (84 per cent of the total land). The remaining peasant households and land were in mountainous areas where ethnic

---

54 Quảng Nam-Đà Nẵng. (1979). Cùng với cả Miền Nam tỉnh ta khẩn trương hoàn thành hợp tác hóa nông nghiệp [Our province, together with southern provinces, hurries to complete collectivisation], *Quảng Nam-Đà Nẵng*, 27 June, p. 1.

55 Năm năm cải tạo xã hội chủ nghĩa đối với nông nghiệp ở Miền Nam [Five years of socialist reform for agriculture in the south], *Nhân Dân*, 29 April 1980, p. 1. Similar praise for collective farming was found in *Quảng Nam-Đà Nẵng* (Our province, together with southern provinces, 27 June 1979, p. 1); and in Thế Đạt (1981), *Nền nông nghiệp Việt Nam từ sau cách mạng tháng Tám năm 1945* [*Vietnamese Agriculture Since the August Revolution 1945*], Hà Nội: NXB Nông Nghiệp, pp. 215–16.

minorities lived.[56] The size of collectives ranged from 200 to 700 hectares and among these were 48 collectives the size of a commune (*xã*). On average, each collective had 421 hectares of agricultural land, 1,542 workers and 762 households.[57] The acceleration of collectivisation in QN-ĐN is illustrated in Table 4.2.

Table 4.2 Seasonal acceleration of collectivisation in QN-ĐN, 1977–79

	Winter–spring 1977–78	Spring–summer 1978	Summer–autumn 1978	Winter–spring 1978–79	Spring–summer 1979	Summer–autumn 1979	End of 1979
No. of collectives	4	4	7	114	132	164	235
Percentage of peasant households	n/a	n/a	n/a	50	57.8	70	92.9

n/a = not available

Sources: Thắng lợi bước đầu của phong trào Hợp tác hóa nông nghiệp [The first victory steps of collectivisation], *Quảng Nam-Đà Nẵng*, 13 May 1978, p. 1; Thành lập xong 98 hợp tác xã sản xuất nông nghiệp [98 agricultural collectives have been established], *Quảng Nam-Đà Nẵng*, 11 October 1978, p. 3; Năm 1979 tỉnh ta căn bản hoàn thành hợp tác hóa nông nghiệp ở các huyện đồng bằng [The midlands of our province have completed collectivisation in 1979], *Quảng Nam-Đà Nẵng*, 17 October 1979, p. 1; Hội nghị Ban cải tạo nông nghiệp tỉnh: Ra sức củng cố HTX để làm tốt vụ sản xuất Đông–Xuân [Provincial Committee for Agricultural Transformation: Strengthening cooperatives to make good in winter–spring production], *Quảng Nam-Đà Nẵng*, 26 December 1979, p. 1.

In Thăng Bình district, as in many other districts in QN-ĐN, collectivisation was rapid and faced no strong peasant resistance. After 'successfully' establishing the experimental collective of Bình Lãnh, in mid-1978 the district's leaders called for the setting up of collectives in other communes. By September 1978, Thăng Bình district had established 10 collectives in five of its 20 communes.[58] By June 1979, Thăng Bình had established 17 collectives in 13 communes, involving 54 per cent of peasant households and 44 per cent of agricultural land; and,

---

56 The midlands of our province, *Quảng Nam-Đà Nẵng*, 17 October 1979, p. 1.

57 Phong trào hợp tác hóa nông nghiệp: Sự kiện và con số [Overview of collectivisation: Events and figures], *Quảng Nam-Đà Nẵng*, 15 June 1983, p. 2.

58 Thăng Bình chuẩn bị xây dựng 10 hợp tác xã [Thăng Bình is about to establish 10 collectives], *Quảng Nam-Đà Nẵng*, 9 September 1978, p. 1.

among these, 10 communes had largely completed collectivisation.[59] By the end of 1979, Thăng Bình's authorities announced the completion of collectivisation, with 36 collectives across its 20 communes.

Bình Định commune in Thăng Bình district completed collectivisation by October 1979, and had two collectives: Bình Định collective no. 1 and Bình Định collective no. 2 (where I did fieldwork and interviews in 2004 and 2005). A former cadre of Bình Định collective no. 2 recalled that, after one month of mobilising peasants, almost all households in the area had joined the collective. Only 20 peasant households declined, most of whose occupants were too old to work.[60]

Thus, by 1979–80, collectives (*hợp tác xã*) were the main farming organisations in QN-ĐN and other provinces of the Central Coast region. A typical collective in QN-ĐN encompassed most or all of a commune and had from 200 to 700 hectares of agricultural land. On average, collectives in QN-ĐN were as large as, or even larger than, typical collectives in the north.[61] For example, the Bình Lãnh commune-sized collective in Thăng Bình district had 1,900 hectares of uncultivated land, 564 hectares of agricultural land and 1,050 households.[62]

QN-ĐN leaders defined collectives as 'socialist agricultural production economic organisations established voluntarily by peasants and under the leadership of the party with the guidance and help of the state'.[63] Peasants over 16 years of age were supposed to do collective work. When participating in the collective, each member had to contribute a share (*cổ phần*) to the collective's assets. Households were allowed to retain part of their land—known as '5 per cent land'—mostly from their garden, for the family economy (*kinh tế gia đình*). All other land and livestock were supposed to be in the collective.[64]

59 Thăng Bình sơ kết hợp tác hóa nông nghiệp, phát động thi đua với HTX Duy Phước, Định Công và Vũ Thắng [A preliminary summing up of collectivisation in Thăng Bình], *Quảng Nam-Đà Nẵng*, 2 June 1979, p. 1.
60 Author's interview, 20 April 2004, Bình Định.
61 Kerkvliet, *The Power of Everyday Politics*, p. 138. The enlarged collectives in the north in 1974 averaged about 200 hectares of land and 350 households.
62 HTX Bình Lãnh từ yếu kém vươn lên tiên tiến [Bình Lãnh collective is moving away from a position of weakness], *Quảng Nam-Đà Nẵng*, 9 June 1979, p. 2.
63 Một số quy định về xây dựng hợp tác xã [Some regulations on establishing collectives], *Quảng Nam-Đà Nẵng*, 26 August 1978, p. 1.
64 Ibid.

QN-ĐN had met the agricultural transformation target set by the VCP's leaders. Collectivisation was completed within one year—even faster than in the north, where collectivisation took two years to complete.[65] In explaining the rapid collectivisation in QN-ĐN, one provincial party leader attributed it to the 'correctness' of the central party's policy, 'the loyalty of local authorities', 'the close relationship between peasants and the party' and the extensive preparations for collectivisation such as organising peasants into labour exchange teams and training a large number of cadres.[66]

My interviews also revealed that local cadres from the provincial to the commune level in QN-ĐN were devoted to the policy. After reunification, provincial cadres were local ex-revolutionaries or returnees from the north. They were familiar with the north's collectivisation model and could apply it to QN-ĐN. People in QN-ĐN had a long history of complying with state policies (phục tùng nhà nước); therefore, after reunification, local cadres and residents tended to comply with the VCP's agricultural transformation policy.[67] Moreover, local cadres were considered more 'bolshevist' and 'fascist' than their counterparts in the Southern Region and were willing to use coercive measures to force peasants to join collectives.[68]

A former vice-chairman of Bình Định collective no. 2 admitted that, at that time, he was loyal to the VCP's agricultural transformation policy because he had been taught that socialism was ideal, and the main task of the new authorities was to transform the old economy and build a new one. He believed collectivisation was the only way to prosperity, justice and the elimination of exploitation. He acknowledged that he had at first trusted the VCP's policy. He reasoned:

> In the war with America, the north carried out collectivisation and supported the south to win the war. As far as I knew, most of the chairpersons of agricultural collectives in the north were women. So, we men could do it.

---

65   Kerkvliet, The Power of Everyday Politics, p. 69.
66   Quảng Nam-Đà Nẵng. (1979). Cuối năm 1978: Ra đời 107 hợp tác xã bao gồm 96,704 nông dân, chiếm 50% số hộ trong tỉnh [By late 1978: 107 cooperatives were established, including 96,704 farmers, accounting for 50 per cent of households in the province], Quảng Nam-Đà Nẵng, 2 December, p. 1.
67   Author's interview, 10 October 2004, Tam Kỳ.
68   Author's interviews, 26 December 2005, Thành Phố, Hồ Chí Minh; 6 October 2004, Bình Định.

… [N]orthern cadres kept us in the dark, too. They did not tell us the shortcomings of collectivisation in the north. Even so, some southern returnees did.[69]

With high expectations for collectivisation and loyalty to the VCP, local cadres were keen to bring peasants into collectives. Some former cadres admitted they had to use various tactics, even 'tricky measures' (*thủ đoạn*) and harsh sanctions, to force peasants to comply. For example, some threatened peasant households who declined to join. Villagers who refused to join faced obstacles in their individual farming efforts and encountered problems with official paperwork, especially that relating to children's schooling and access to health care and state goods.[70] These measures were quite similar to those used in the north during 1959–61.[71]

In theory, each collective was built on three principles: voluntary membership, mutual benefit and democratic governance. In reality, peasants were coerced into joining the collective. A former chairman of Bình Lãnh collective recalled:

> The first principle of collectivisation was coercion. Livestock [*trâu bò*] and land were all collectivised. Right before establishing the collective, application forms were sent to ask peasants to sign. If someone declined to join, his land was replaced with barren land far away so that peasants joined out of fear.[72]

In contrast to previous research suggesting the rapid collectivisation in the Central Coast region was due to peasants living in unfavourable conditions and 'hoping for a better life', my interviews show that many peasants joined collectives out of fear.[73] Many villagers who joined recalled 'being coerced' (*bị bắt buộc*), 'fearing isolation' (*sợ bị cô lập*) and 'fearing disadvantage' (*sợ bị thua thiệt*). Some also said they joined out of 'ignorance' (*không biết*), because they were 'just following others' (*họ vào thì mình vào*) and because 'being poor together was okay' (*nghèo thì nghèo chung*).[74] Asked whether he volunteered to participate, one 60-year-old man in Thanh Yên village of Bình Định commune, Thăng Bình district, recalled:

---

69 Author's interview, 20 April 2004, Bình Định.
70 Author's interview, 21 October 2004, Bình Lãnh.
71 See Kerkvliet, *The Power of Everyday Politics*, p. 71.
72 Author's interview, 21 October 2004, Bình Lãnh.
73 Quang Truong, Agricultural collectivization, p. 207.
74 Author's interviews, October 2004, Thanh Yên and Hiền Lộc.

If we did not participate, we had to endure a lot of disadvantages; we could not keep our land but were given bad land far away. During the collectivisation campaign, local cadres warned that if we did not join, our cows and buffaloes would not be allowed to graze on, even go through, collective fields. Moreover, if we did not join, we would be isolated from other people; we could not buy goods from the state; our children would not have access to education and other things.[75]

A recent unpublished essay by a well-known journalist in QN-ĐN reported that, when the Bình Lãnh, Duy Xuyên and Hòa Tiến 1 pilot collectives were established, peasants in other areas of QN-ĐN worried (*lo lắng*), doubted (*nghi ngờ*) and feared (*sợ hãi*) that collectivisation would reach them.

Many explanations and discrediting of rumours that collectivisation in the north had produced bad consequences failed to stop peasants in many areas of QN-ĐN slaughtering or selling their animals before collectivisation, destroying plants and selling their agricultural machinery.[76]

Villagers in Thanh Yên village, Bình Định commune, admitted that many people slaughtered or sold their draught animals to avoid collectivising them.[77] However, although many peasants doubted the benefits of collective farming, most joined.

The Mekong Delta

As previously mentioned, in 1977, authorities in the Mekong Delta had great trouble establishing experimental commune-sized collectives. Hence, in 1978, they scaled back their expectations and concentrated on setting up hamlet-sized ones and then production units (*tập đoàn sản xuất*) with between 30 and 50 hectares of land. At the conference on agricultural transformation in the south held in Cửu Long province in April 1979, national leaders claimed that production

---

75 Author's interview, 17 October 2005, Thanh Yên. It was said that peasants who joined collectives received purchasing books (*sổ mua bán*) that enabled them to buy fuel, soap, salt, clothes and other goods in state shops. Non-collective members could not access these goods (Author's interview, 7 October 2005, Thanh Yên).

76 L. K. (1990), Từ quá khứ đến hiện tại: Mười lăm năm ấy [From past to present: Over the past 10 years], Unpublished essay. I was given this article when I interviewed the author, a former *Quảng Nam-Đà Nẵng* journalist, on 20 October 2005.

77 Author's interviews, 1–30 October 2005, Bình Định.

units were the most suitable collective organisation for the Southern Region as a whole and for the Mekong Delta in particular. Therefore, they called for the region to accelerate collectivisation in the form of production units instead of through collectives. However, the national leaders still wanted to experiment with collectives and hoped that many more could eventually be set up by consolidating well-established production units.[78]

In 1979, in response to the VCP leaders' policy, An Giang and other provinces in the Southern Region (and in the Mekong Delta and South-Eastern Region) accelerated the formation of production units. According to a report of the Central Committee for Southern Agricultural Transformation (BCTNNMN), by November 1979, the Southern Region had established 13,178 production units and 272 pilot collectives, accounting for 33.5 per cent of peasant households and 26.9 per cent of agricultural land.[79] However, according to the *Nhân Dân* newspaper in April 1980, few of these collectives operated well; many failed to show 'the superiority of new production relations' and failed as 'an appropriate form of collective'.[80] Moreover, only 7,000 production units in the Southern Region actually farmed collectively (*làm ăn chung*), and even these faced many difficulties. The remaining production units had not yet started to farm or had started but failed. Production unit members 'still did not feel secure' (*vẫn chưa an tâm*), even in some of the well-performing units. Some units deviated from their production schedule (*tiêu cực trong sản xuất*), illegally giving 'blank contracts' to households (*khoán trắng cho hộ*), and were verging on collapse (*sắp tan rã*).[81]

---

78 BCTNNMN (1978), *Thông báo về cuộc họp từ ngày 22–24 tháng 10 năm 1979 của Ban cải tạo nông nghiệp Miền Nam* [*Report of Central Committee for Agricultural Transformation in the South on 22–24 October 1979 Meeting*], 5 November, Hồ Chí Minh: Ban Cải Tạo Nông Nghiệp Miền Nam, p. 8.

79 Ibid.

80 Five years of socialist reform, *Nhân Dân*, 29 April 1980, p. 1. This article also shows that most of the pilot collectives in the Southern Region were in Sông Bé province (152 collectives) and Tiền Giang (70 collectives).

81 BCTNNMN (1979), *Thông tri về việc kịp thời và ra sức củng cố các tập đoàn sản xuất nông nghiệp* [*Announcement on Doing the Best to Improve Production Units*], 1 November, Hồ Chí Minh: Ban Cải Tạo Nông Nghiệp Miền Nam, pp. 1–2.

A typical example is Minh Hải province in the Mekong Delta, where collectivisation accelerated extensively in 1979. Within a year, the province had 1,114 production units, involving 45.8 per cent of households and 36 per cent of agricultural land (see Table 4.3).

Only 300 of 1,114 production units were actually farming collectively (*tập đoàn ăn chia*), and only 130 of these 300 units had socialist qualities (*tập đoàn theo đúng tính chất xã hội chủ nghĩa*)—that is, they produced collectively and distributed their output according to the work-points members earned. The remaining 170 production units only farmed semi-collectively. This means that, while some farming was done collectively, individual households privately cultivated their own land or part of the production unit's land.[82]

Table 4.3 Accelerating collectivisation in Minh Hải province, 1979

Date	Tasks	No. of production units
January 1979	Experimental pilot production units	3
March 1979	Extending pilot production units	12
April 1979	Accelerating collectivisation	100
May 1979	Accelerating collectivisation	500
June 1979	Accelerating collectivisation	800
August 1979	Accelerating collectivisation	1,081
October 1979	Accelerating collectivisation	1,114

Source: Ban Cải Tạo Nông Nghiệp Minh Hải (BCTNNMH). (1979). Dự thảo báo cáo: Nhận định, đánh giá tình hình cải tạo nông nghiệp thời gian qua ở Minh Hải [A Draft Report: Evaluation of Agricultural Transformation in Minh Hải], 13 November. Minh Hải: Ban Cải Tạo Nông Nghiệp tỉnh Minh Hải.

In response to the poor results of collectivisation in the Southern Region, party leaders in November 1979 instructed everyone to 'try their best to strengthen production unit organisations'.[83] The national leaders also lowered their expectations and called for the acceleration of collectivisation in a 'positive and firm way' (*phương châm tích cực và vững chắc*) instead of the 'urgent way' (*khẩn trương*) advocated in

---

82   Ban Cải Tạo Nông Nghiệp Minh Hải (BCTNNMH). (1979). *Dự thảo báo cáo: Nhận định, đánh giá tình hình cải tạo nông nghiệp thời gian qua ở Minh Hải [A Draft Report: Evaluation of Agricultural Transformation in Minh Hải]*, 13 November. Minh Hải: Ban Cải Tạo Nông Nghiệp tỉnh Minh Hải..

83   BCTNNMN, *Announcement on Doing the Best to Improve Production Units*, pp. 1–2.

previous policies. They also instructed the Southern Region to focus on solidifying (*củng cố*) existing collective organisations rather than accelerating the formation of new ones. In particular, the region was to put much more emphasis on improving the quality of local cadres and creating the 'necessary conditions' for them to avoid having to carry out collectivisation in a subjective, hasty and coercive way, which was perceived to be harming production and living standards.[84]

Despite the VCP leaders' efforts, by early 1980, more than two-thirds of the production units in the Southern Region had collapsed. For example, of the 2,653 production units established in Hậu Giang province in 1979, 'there were no more than 100' that could 'stand firm'.[85] By the end of 1980, there were only 3,729 production units and 137 collectives remaining in the Southern Region.[86] These collective organisations accounted for only 8 per cent of peasant households and 6 per cent of land.[87]

VCP leaders often blamed local cadres for the poor results, accusing those in the Southern Region of being 'simpleminded' and 'hasty', suggesting they 'propagandised and mobilised the masses inadequately' (*thiếu tuyên truyền và vận động quần chúng*), 'coerced the masses' and had committed 'shortcomings' in management.[88] The BCTNNMN's report revealed that many local-level cadres in the Southern Region did not grasp fully the content of collectivisation policy and had not studied it thoroughly. They were therefore unable to explain the policy to lower-level cadres and the masses and erred in their instructions, making collectivisation even harder to implement.[89] In evaluating the obstacles to slow collectivisation in the Southern Region, VCP leaders complained that:

---

84 ĐCSVN (2004), Nghị quyết hội nghị lần thứ 6 Ban chấp hành Trung ương Đảng khóa IV [Resolution of the 6th Plenum of the Fourth Party Central Committee], in ĐCVSN, *Văn Kiện Đảng Toàn Tập: Tập 40, 1979* [*Party Document: Volume 40, 1979*], Hà Nội: NXB Chính Trị Quốc Gia, p. 362.

85 Phan Quang, *The Mekong Delta*, p. 83.

86 Vũ Oanh (1984), *Hoàn thành điều chỉnh ruộng đất, đẩy mạnh cải tạo xã hội chủ nghĩa đối với nông nghiệp các tỉnh Nam Bộ* [*Completing Land Redistribution and Speeding Up Agricultural Transformation in the Southern Region*], Hà Nội: NXB Sự Thật, p. 11.

87 Lê Thanh Nghị (1981), *Cải tiến công tác khoán sản phẩm để thúc đẩy sản xuất, củng cố HTX nông nghiệp* [*Improving the Product Contract to Solidify Collectives*], Hà Nội: NXB Sự Thật, p. 33.

88 Ngo Vinh Long, Some aspects of cooperativization in the Mekong Delta, p. 166; BCTNNMN, *Announcement on Doing the Best to Improve Production Units*, p. 2.

89 BCTNNMN, *Report of Central Committee for Agricultural Transformation*, p. 6.

Local cadres and party members including key cadres have not yet sympathised with the agricultural transformation revolution [*chưa cảm tình với cách mạng cải tạo nông nghiệp*]. They still neglect [*thờ ơ*] and do not support it. They stand outside and leave the task of collectivisation to other specialised departments. Besides this, some negative cadres who were pursuing their own interests did not want to implement collectivisation. When it went smoothly, they were silent but when collectivisation went badly, they criticised it by amplifying its shortcomings and exacerbating the situation. They tolerated 'bad elements' who harmed the process.[90]

An Giang province

After reunification, local authorities in many parts of An Giang province faced numerous difficulties in consolidating their power, with a crucial problem being a shortage of cadres to fill positions of local authority. Additionally, southerners returning from the north seldom worked at the local level. So, to find new cadres, local authorities had to recruit people who were not familiar and did not have any experience with the VCP's agrarian policies, especially collectivisation.

Despite these difficulties, by the end of 1979, An Giang had established 308 production units, six pilot collectives and 55 machinery units (*tập đoàn máy*); collective organisations accounted for about 5 per cent of agricultural land and 7 per cent of peasant households.[91] However, the majority of these production units were classified as weak and of inadequate quality (*chưa đúng tính chất*). For example, Phú Tân district had established six production units in 1979 but only two were involved in collective farming. Likewise, only nine of 94 production units in Châu Thành district had 'socialist characteristics'.[92] Some production units faced difficulties due to peasants' resistance and were dismantled a few months after being established. In some districts where a majority of people were of Khmer ethnicity, such as Tịnh Biên and Tri Tôn, no production units had been established.[93]

---

90   Ibid., p. 5.
91   An Giang vững vàng đi tới [An Giang is doing well], *An Giang*, 6 January 1980, p. 1. The proportions of land and peasant households belonging to collective organisations are based on my own calculations.
92   Collectivisation continues to progress, *An Giang*, 7 June 1981, p. 2.
93   Author's interviews with provincial cadres, 31 May 2005. An Giang's local archives remained almost silent on this matter so I cannot know exactly how many production units were dismantled in 1979.

Faced with great difficulties in extending the number of production units and making them function as collectives, An Giang's leaders in 1980–81 put more effort into consolidating existing production units rather than rapidly creating more (see more detail in the next section). As a result, collectivisation during this period stagnated. At the end of 1980, An Giang had 317 production units, six collectives, 1,584 production solidarity teams and 64 machinery units.[94] By the end of 1981, An Giang still had only six collectives, while the number of production units had risen by 40 to 357. These collective organisations occupied about 20,675 hectares of agricultural land, a mere 8.5 per cent of the total, and involved 10 per cent of peasant households.

In Chợ Mới district in An Giang, as well as experimenting with pilot production units, authorities cautiously extended their number. By the end of 1979, the district had 19 production units; however, most were weak and 'infirm' (không vững chắc) and the cadres managing them were described as 'confused' (lúng túng).[95] According to a former Chợ Mới official, because of difficulties in extending the production units, authorities emphasised solidifying existing units, meaning that, in 1980, collectivisation stagnated.[96] Only a few units were established in the district in 1980.[97] Thus, by the end of 1980, Chợ Mới had established only 21 production units, which accounted for about 5.7 per cent of peasant households and 4.7 per cent of agricultural land.[98] By 1981, the district had established 19 additional production units. Therefore, during 1979–81, Chợ Mới district established 40 production units, which accounted for only 10 per cent of the peasant households and 8.5 per cent of agricultural land.[99]

---

94   Trong tháng 12, 1980 tỉnh phát triển thêm được 14 tập đoàn sản xuất [In December 1980, the province established 14 more production units], *An Giang*, 11 January 1981, p. 2.

95   Phong trào hợp tác nông nghiệp ở An Giang từng bước được củng cố đi lên [Collectivisation in An Giang has progressed], *An Giang*, 18 November 1981, p. 1; BCHDBHCM, *The History of Chợ Mới Party Cell*, pp. 174–5.

96   Author's interview, 17 August 2005, Long Điền B.

97   Report from a conference announcing the completion of collectivisation in Chợ Mới, *An Giang*, 15 April 1985, p. 1.

98   My calculation is based on figures reported in *An Giang* (ibid.); and in Chi Cục Thống Kê huyện Chợ Mới [hereinafter CCTKCM] (1984), *Niêm giám thống kê 1976–1984 huyện Chợ Mới tỉnh An Giang* [*Chợ Mới District, An Giang Province Statistical Year Book, 1976–1984*], Chợ Mới: Chi Cục Thống Kê huyện Chợ Mới, p. 43.

99   CCTKCM, *Chợ Mới District, An Giang Province Statistical Year Book*.

In Long Điền B in 1980, after establishing two pilot production units in 1979, and with help from district leaders, authorities established two more production units, located near the previous ones. District and commune officials strove to make these four production units work properly and act as exemplary cases. A former Long Điền B commune official claimed that, because of such efforts, commune authorities did not extend collectivisation further. Therefore, between 1979 and 1981, Long Điền B established only four production units, which accounted for a modest proportion of both agricultural land and peasant households in the commune.

In general, collectivisation in An Giang and other provinces in the Mekong Delta met with substantial difficulty and proceeded very slowly. Regional collectivisation accounted for only a small proportion of land and peasants (less than 10 per cent), which fell far short of the VCP's target of completing collectivisation by 1980. The VCP leaders attributed the slow collectivisation in the Southern Region to local cadres, who, they claimed, were 'hesitant [*do dự*], tentative [*chần chừ*] and undetermined [*thiếu kiên quyết*] in carrying out collectivisation, and too relaxed about agricultural transformation [*buông lỏng cải tạo*]'. Some were accused of manipulating the VCP's 'positive and firm principles' to delay collectivisation.[100]

Similarly, when collectivisation was slow and difficult in An Giang in 1980 and early 1981, provincial leaders shifted all blame to lower-level local cadres. They said the local authorities, especially in the communes, lacked 'determined, integrated and concerted leadership'. 'Some local leaders were lax about agricultural transformation'; local cadres were 'inadequate and weak' (*thiếu và yếu*), so

> the capacity of local agricultural transformation bodies did not match with their function and obligations.
>
> … [S]ome cadres had not grasped or intentionally misunderstood the content of the VCP's policy on agricultural transformation.
>
> … They resorted to the VCP's principle of firm collectivisation and voluntary membership to maintain individual farming.

---

100 ĐCSVN (2005), Chị thị của Ban bí thư số 93/CT-TW (ngày 30 tháng 6 năm 1980) [Directive of the Secretariat No. 93/CT-TW (30 June 1980)], in ĐCSVN, *Văn Kiện Đảng Toàn Tập: Tập 41, 1980* [*Party Document: Volume 41, 1980*], Hà Nội: NXB Chí Trị Quốc Gia, p. 204.

Finally, at the production unit level,

negativism occurred in some management boards.

… Some production units achieved poor outcomes so [people's] living conditions had not been improved.

… [All of which] raised doubts and undermined peasants' confidence in VCP's agricultural transformation policy.[101]

A former cadre of An Giang's Committee for Agricultural Transformation (Ban Cải Tạo Nông Nghiệp, or BCTNN) admitted that, despite trying to secure their positions, local cadres were less devoted to collectivisation because 'the policy was at odds with peasants' sentiments' (*không hợp lòng dân*). Some cadres therefore 'let the process of agricultural transformation drift'.[102]

Sharing a similar view, another cadre of An Giang's BCTNN added:

Implementing collectivisation in the Mekong Delta seemed less harsh than in the Central Coast because local authorities tended to use persuasion and less coercion to force peasants to participate in collective organisations. Therefore, agricultural transformation in the Mekong Delta had not been carried out completely [*không triệt để*]. Collectivisation went slowly because of peasants' strong reaction and cadres' hesitance.[103]

Many villagers in Long Điền B recalled that before collectivisation they were relatively well off and had enjoyed sufficient livelihoods (*sung túc*). Even agricultural workers had been able to lead a comfortable life (*sống thoải mái*). This explains why most peasants, even those who were poor, did not want to join production units with work-points systems in which they would earn only a little (*không có ăn*). Some said they did not like collective farming because it constrained the freedom (*bị gò bó*) they had previously enjoyed under individual farming.[104] One man in Long Điền B commune commented:

People in the Central Coast and in the north were used to living in poverty [*sống kham khổ quen rồi*], so they could accept collectives, but people in this region had become used to enjoying a sufficient and free

---

101 Collectivisation continues to progress, *An Giang*, 7 June 1982, p. 2.
102 Author's interview, 6 June 2005, Long Xuyên.
103 Ibid.
104 Author's interviews, June–July and 5 August 2005, Long Điền B.

life. They did not like life in the collectives with little freedom. Peasants could not be like factory workers—the bell rings and they march off to work. Peasants here wanted more time to enjoy breakfast, coffee or to take care of their children and animals. Moreover, peasants here did not like joining production units because they did not see any immediate and visible benefit in collective farming [*không thấy lợi trước mắt*].

Some landless and land-poor households in Long Điền B also refused to join production units. One landless man in the commune argued that earnings from collective farming were less than the income from wage labour. In addition, people in production units received their produce at the end of the season, while independent labourers received wages on a daily basis.[105] Another landless man who had previously sympathised with the revolution but who refused to undertake collective farming recalled production unit cadres inviting him to join the unit in his hamlet. If he did not join, the cadre warned, and if later he faced hunger, the production unit would not lend him rice, and there would be no land on which to bury his body when he died. Regardless of what cadres threatened, the man refused to join. He reasoned that under collective farming he would earn much less than from his current job raising pigs, gleaning leftover paddy in the fields and labouring for wages. He laughed and added, 'ultimately, not me but members of production units came to borrow my rice'.[106]

Peasants in the Mekong Delta tended to resist collectivisation more strongly than their counterparts in the Central Coast. In some parts of the delta peasants boycotted or organised strikes against collective farming and even threatened to assassinate—and, in some cases, actually did assassinate—officials. According to Vo Nhan Tri, peasants in some parts of the Mekong Delta 'refused to harvest crops in time, abandoned large stretches of land, slaughtered livestock, destroyed fruit trees, sold machines and farm implements before joining the production units'.[107] Similarly, a report from the BCTNNMN revealed:

> In some locations in Long An province collectivisation was so stressful that peasants, incited by the enemy, formed groups to demand their departure from production units, protested against collective farming and rallied support for individual farming [*chống đối làm ăn chung, ủng*

---

105 Author's interview, 17 August 2005, Long Điền B.
106 Author's interview, 23 June 2005, Long Điền B.
107 Vo Nhan Tri, *Vietnam's Economic Policy Since 1975*, p. 79.

hộ làm ăn riêng lẻ] … [Moreover,] taking advantage of the difficulties of collectivisation, counter-revolutionaries and bad elements conducted sabotage activities. They carried out psychological warfare such as distorting agricultural transformation policy, sabotaging production, assassinating local key cadres and inciting the masses to strike against the government, destroying production units' seed stores, beating local cadres and harvesting collective rice illegally. Some tried to enter the managerial boards of production units and collectives and so on.[108]

Conclusion

After the country's reunification in 1975, VCP leaders put great effort into imposing the north's collective models on the south and aimed to complete collectivisation there by 1980. However, the project encountered difficulties, especially in the Mekong Delta. Authorities in QN-ĐN achieved the central government's target to collectivise farming in that province by 1980, but, in contrast, authorities in An Giang and elsewhere in the Mekong Delta did not. Collectivisation in the delta accounted for less than 10 per cent of land and peasant households in 1980. Only in the mid-1980s was collectivisation deemed accomplished, thanks in part to policy modifications to accommodate villagers' concerns.

There were two major reasons for the differences in the outcome of collectivisation between these two places. First, the capacity of local authorities was greater in QN-ĐN than in An Giang. Local authorities in QN-ĐN had more experience with the VCP's northern collectivisation and were more loyal to its policy of socialist transformation of agriculture than their counterparts in An Giang. Provincial authorities in QN-ĐN therefore carried out collectivisation more aggressively; they used stronger coercive measures—similar to those used in the north in the early 1960s—to force villagers into collectives. They collectivised all land, draught animals and other peasant means of production simultaneously, tightly restricted private farming and handicapped non-members. They even used strict preemptive measures to prevent peasants from slaughtering animals or restricting villagers' mobility before collectivisation. In contrast, local cadres in An Giang had weaker commitment to agricultural transformation. To secure their

---

108 BCTNNMN, *Report of Central Committee for Agricultural Transformation*, p. 4.

positions, many had to comply with national policies, but they did so unenthusiastically. In general, when faced with strong peasant resistance, many local cadres were reluctant to force policy compliance; often they modified policies or let the process drift to accommodate peasants' concerns.

Second, peasants' noncompliance was stronger in An Giang in the Mekong Delta than in QN-ĐN in the Central Coast. The consequences of war in QN-ĐN had been so severe they rendered most villagers poor. These villagers' main concerns were to do with subsistence and survival. They were living in extremely difficult conditions within corporate communities and had few outside employment opportunities.[109] Their behaviour was focused on securing their own safety, subsistence and survival. The strong local authorities who were insisting on implementing state policies were ready to impose heavy sanctions on those who did not comply. To avoid suffering disadvantages, many peasants decided to join collectives even though they did not believe in the benefits of collective farming. Authorities in QN-ĐN had earned a fair degree of legitimacy thanks to ending the war and carrying out previous land reforms, which made peasants more inclined to comply with official policy. Thus, authorities in QN-ĐN were able to complete collectivisation within a year—even faster than collectivisation in the north in the early 1960s. In contrast to previous scholars who attributed swift collectivisation in the Central Coast region to peasants' preference for collective farming as a means of coping with their difficult lives, I found that many villagers in Thanh Yen and Hien Loc initially disliked and did not trust such farming methods, but they decided to join collectives to avoid the disadvantages of not belonging.[110]

Meanwhile, peasants in An Giang were better off than those in QN-ĐN and lived in diverse socioeconomic structures. Moreover, weaker and less legitimate local authorities who were hesitant about and incompetent in forcefully carrying out socialist agricultural transformation enlarged the scope for villagers to evade the state's policies. In such conditions, An Giang villagers were able to resist or evade agrarian projects that

---

109 Scott, J. C. (1977), *The Moral Economy of the Peasant: Rebellion and Subsistence in Southeast Asia*, New Haven, CT: Yale University Press.
110 See Quang Truong, Agricultural collectivization.

they considered were unattractive or unprofitable. They had more economic power and more options to resist collective farming, which they saw as inferior to their previous farming method.

In short, collectivisation in An Giang and the wider Mekong Delta faced more difficulties than in QN-ĐN in the Central Coast region because of stronger peasant resistance and weaker capacity and commitment of local cadres. In other words, the extent of collectivisation depended largely on the political and socioeconomic conditions of each region.

5

Local politics and the performance of collective farming under the work-points system, 1978–81

Introduction

Despite Vietnamese Communist Party (VCP) leaders and local authorities putting great effort into establishing and strengthening collective organisations, collective farming had failed to show its superiority over individual farming in southern Vietnam. Similar to what Kerkvliet's study found for the north, in the south, due to a lack of political conditions conducive to durability, collective farming also became a site for struggle between peasants and local cadres and between local people and higher authorities over the governance of collectives, the means of production (land, labour and other resources) and distribution of produce.[1]

Under the work-points system (1978–81), these struggles caused major difficulties not only in the Mekong Delta, but also in the Central Coast region. Villagers in the Mekong Delta tried their best to evade collective farming; it was common for them to join a collective but not actually participate in its work. Meanwhile, although villagers in the Central Coast seemed to comply with the system, they tried their best to maximise their work-points rather than production. In addition, local

---

1   Kerkvliet, *The Power of Everyday Politics*, p. 20.

cadres in both regions often acted contrary to the expectations of VCP leaders and villagers. They often took advantage of their positions to embezzle resources and mismanage collectives. Despite the authorities' numerous campaigns to correct and crack down on such 'bad behaviour', and even attempts to modify national policies to accommodate local concerns, these problems did not disappear, but seemed to increase.

This chapter will examine local politics and compare the forms and magnitude of peasants' everyday politics and local cadres' reactions and malpractice in response to collective farming and other national agrarian policies. It will also examine the extent to which the everyday practices of local cadres and peasants affected the outcome of collectivisation in both regions and how they contributed to the modifications of the VCP's agrarian policy.

Local politics in Quảng Nam-Đà Nẵng (QN-ĐN), Central Coast region

Peasants' everyday politics in QN-ĐN's collectives

In theory, collectives were established according to the principles of voluntary membership, mutual benefit and democratic management. According to public pronouncements, peasants were the 'masters of the collective'. During the collectivisation campaigns, local authorities in QN-ĐN often asserted that 'the collective was the home and its members were the masters' (*hợp tác xã là nhà, xã viên là chủ*).[2] However, may peasants claimed they did not join collectives voluntarily, but were coerced into doing so. Most peasants preferred individual farming to pooling their resources. They had doubts about collective farming methods and believed the collectives belonged to the state. Many worried that the collectives were being managed poorly and that much of what the collective produced would be stolen. Therefore, collectives became sites of conflict and struggle between peasants, cadres and the state, and even among peasants themselves. In struggling for their livelihoods, peasants tended to do whatever favoured their own immediate interests, which was often at odds with the interests of the

---

2   Vai trò của đảng viên trong đội sản xuất [The role of party members in production brigades], *Quảng Nam-Đà Nẵng*, 28 June 1978, p. 2.

collective and the state. The next section classifies and examines the nature of the peasants' everyday politics in the collectives in QN-ĐN during the period of the work-points system (1978–81).

Optimising work-points rather than the quality of production

Peasants in QN-ĐN were relatively poor and few had any economic options outside the collective, so earning work-points was important to their income. The larger the number of work-points they accrued, the more paddy they were supposed to receive.

Many poor peasants in Hiển Lộc and Thanh Yên villages claimed they had to fight for work-points (*tranh giành công điểm*), working as fast as possible to acquire more points. A widow with four small children recalled:

> I took advantage of any opportunity to get more work-points. As soon as people harvested, I jumped to hoe the corner of the plot in order to take over ploughing it. If I did not do so, others would. As soon as I had finished, I changed to another plot. My little girl, aged 13, also pulled up rice seedlings to get points. If an adult got 10 points a day, she got five. At that time, I did not have time to rest.[3]

Similarly, another widow said:

> I had to struggle to get work-points [*phấn đấu để lấy điểm*]. I was the only labourer in my family. We lacked labour because of [the] loss of [working-aged] men … due to wars. So, we had to work hard by day and night to get work-points.[4]

Asked why peasants struggled to get work-points, an elderly man in Thanh Yên village responded:

> Today we can seek other jobs in Saigon or Danang city, but at that time, if we did not work, we would die of hunger. So, we even had to do a job that earned only a very few work-points.[5]

The *Quảng Nam-Đà Nẵng* newspaper reported in December 1978 that, despite 90–95 per cent of peasants participating in collective work, many focused only on earning work-points. It called the phenomenon 'work-point syndrome' and 'the doctrine of work-points' (*chủ nghĩa công điểm*), which it said had started to 'encroach on peasants' awareness

---

3   Author's interview, 15 October 2005, Hiển Lộc.
4   Ibid.
5   Author's interview, 31 October 2005, Thanh Yên.

of collective mastery'. For example, when peasants were requested to attend public meetings or do public work, they asked whether such things would bring them any work-points. This made peasants very choosy, and they refused to do tasks that would earn only a few work-points, preferring tasks worth more points. 'They were only concerned with the work-points … without caring about what the brigade leaders and others did and expected', the newspaper lamented.[6]

Villagers in Thanh Yên village remembered doing their collective work carelessly and deceitfully (*làm gian làm dối*) to earn as many work-points as possible. For example, when ploughing, they would plough one row and skip the next (*cày một đàng bỏ đàng*). When transplanting, they planted densely at the edges of the plot but sparsely in the centre. Similarly, when weeding, 'they did it carefully on the edges but carelessly in the centre' because an inspector could more readily see the edges.[7] Peasants could earn work-points by selling their manure to the collective, so, to increase its volume, they mixed manure with other easy-to-find ingredients such as rice stubble, soil and leaves.[8] When they were required to carry manure to distant plots, peasants would pour out some of it close to their home or in bushes along the way. In this way, plots close to the village received considerable levels of manure, while more distant plots received little. People also spread the manure unevenly, so that some areas received too much manure while others received nothing (*chỗ có chỗ không*). In some cases, people did not spread manure at all before ploughing and raking.[9] When pulling up seedlings, workers tried to increase the number of seedling bundles by making them smaller than normal in order to maximise their work-points.[10]

A woman in Hiển Lộc village recalled cadres devising a new method of transplanting: putting seedlings in lines. People resisted, however, because this method was slow. 'We transplanted only a few rows half a day at a time. We complained a lot because transplanting like that meant fewer work-points. Finally, they [cadres] gave up the technique', the woman said.[11]

---

6   Nhìn vào đồng ruộng tập thể: Chủ nghĩa công điểm [Looking at collective fields: 'Work-pointism'], *Quảng Nam-Đà Nẵng*, 13 December 1978, p. 2.
7   Author's interview, 5 October 2004, Thanh Yên.
8   Author's interview, 9 December 2005, Thanh Yên.
9   Author's interview, 20 October 2005, Hiển Lộc.
10   Author's interview, 17 December 2005, Hiển Lộc.
11   Author's interview, 15 October 2005, Hiển Lộc.

A newspaper report from December 1978 stated:

> Some peasants only pursue their own interests, so they do collective work deceitfully and carelessly [*làm gian dối, làm ẩu*], never ensured work quality, nor did they comply with technical procedures. People preferred to do easy jobs and refused to do hard ones … they did not harmonise [the] interests of [the] individual, the collective and the state.[12]

The article also attributed such problems to inadequate education.

However, when people were asked why they were so careless about their collective work, many responded that it was 'in order to get as many work-points as possible' or, as one person summed it up: 'work honestly, eat gruel; work deceitfully, eat rice [*làm thật ăn cháo, làm láo ăn cơm*].' People believed that those who tried to do collective work properly and honestly would earn fewer work-points than those who did things carelessly and deceitfully. Such practices in QN-ĐN—aimed at accumulating the maximum number of work-points rather than maximising production—were similar to those in the north, studied by Kerkvliet. For example, an expression similar to the one cited above— 'work well, eat gruel; work deceitfully, eat rice' (*làm tốt ăn cháo, làm láo ăn cơm*)—was also popular in northern collectives in the late 1970s.[13]

One village woman tried to justify people's behaviour by arguing that

> everyone had to try and make a living. If you traded, you sought a profit; if you worked for the collective, you had to try to get work-points; so, people did collective work carelessly in order to get as many work-points as possible.[14]

Lack of incentive and 'neglect of common property' (*cha chung không ai khóc*)

That peasants carried out collective work carelessly and deceitfully reflected not only their strategies to maximise work-points, but also their disillusionment with the governance of collective farming. A man in Hiền Lộc village recalled that people were disappointed because

---

12   Tổ chức lại sản xuất, phân công lại lao động nhằm phát triển và mở rộng lại ngành nghề sản xuất và kinh tế gia đình trong hợp tác xã nông nghiệp trên địa bàn huyện [Reorganising production and labour to facilitate development of handicrafts and household economy in the district], *Quảng Nam-Đà Nẵng*, 13 December 1978, p. 1.
13   See Kerkvliet, *The Power of Everyday Politics*, p. 163.
14   Author's interview, 21 October 2005, Hiền Lộc.

they received few rewards for their efforts; therefore, they did collective work badly, just going through the motions of working and tried to complete the job as soon as possible in order to go home. He said when collective work was assigned to a group, 'they often dragged their feet so that, by 7–8 am, they hadn't even started yet. Those who arrived early did not work until the whole group had come.'[15]

Some peasants did not want to work hard because they realised that, no matter how much effort they put in, they would not get a significant extra reward. A man in Thanh Yên explained:

> The collective took all of what we produced; the collective paid us about 0.5 kilograms a workday [10 work-points] and took all the remainder. So, peasants just went through [the] motions of working.[16]

An elderly man in the village had a similar comment:

> No matter how hard you worked, you could only get 10 points a day at maximum. No matter how industriously you worked, the produce belonged to the collective. So, there was not much difference between industrious workers or lazy workers. We worked without any incentive [làm không có động cơ].[17]

Some peasants were initially eager to fight for work-points but, when they received little reward, they were disappointed and did not want to go to work or laboured unenthusiastically.[18] A brigade leader in Hiền Lộc village commented on the decreased income in the Bình Lãnh collective:

> The living conditions of people went down dramatically. At first, people received 3 kilograms per workday; that went down to 1.5 kilograms [in 1978]. When I called the people to transplant, some refused to work; they complained that they had previously received 3 kilograms but now only 1.5 kilograms per workday and wanted to know why.[19]

---

15 Author's interview, 19 October 2005, Hiền Lộc.
16 Author's interview, 5 October 2004, Thanh Yên.
17 Author's interview, 20 October 2005, Thanh Yên.
18 Author's interview, 15 October 2005, Hiền Lộc.
19 Author's interview, 14 October 2005, Hiền Lộc. In response to low peasant participation in collective work, Bình Lãnh collective started to increase sanctions by setting the number of compulsory workdays for peasants and restricting or forbidding peasants from doing non-collective work (Hợp tác xã Bình Lãnh vượt khó khăn giành thắng lợi bước đầu [Bình Lãnh collective overcame difficulties and gained first good results], Quảng Nam-Đà Nẵng, 13 May 1978, p. 2; Chi bộ Bình Lãnh lãnh đạo xây dựng hợp tác xã nông nghiệp [Bình Lãnh party cell leads building of the agricultural collective], Quảng Nam-Đà Nẵng, 14 June 1978, p. 2).

Another common peasant practice in QN-ĐN collectives was not caring for collective property. The *Quảng Nam-Đà Nẵng* newspaper noted in June 1979 that peasants in Điện Bàn district 'considered the collective belonged to the managerial board and brigade leaders, so they were not active in protecting collective properties from loss or damage'.[20] Likewise, an elderly man in Hiền Lộc village said:

> Working in the collective, Mr Brigade Leader [*Ông đội trưởng*] was in charge of everything while we were only concerned with work-points. Today, I have the red book [the certificate of the right to use land] for my land, so I have made the edges of my plots straight and have levelled the surface because I am the owner of the land. But at that time we did not control the land. If I saw edges of the plot broken, at most I might inform the brigade leader. If he gave me some work-points to repair it, I did [so]. Otherwise, I did not. But if that plot was ours, we would do it.[21]

Another man recalled:

> Working for the collective, we did not need to think; when finishing work, I went to bed without worrying about tomorrow. We let the brigade leader worry about matters. When he asked me to plough, I ploughed. Only later when I worked for myself did I plan everything.[22]

Interest in the family economy and the plundering of collective resources

In addition to collective farming, peasants were allowed to farm individually on their garden land—known as the '5 per cent land'—which the collective set aside for the peasant family economy. In the lowlands and midlands, households were able to retain about 500 square metres for this purpose, while in the highlands, it was about 750–1,000 square metres.[23] Farming on garden land became a central part of peasants' family economy because there were few economic options outside the collective. Peasants in Bình Định collective no. 2 and Bình Lãnh collective often grew sweet potatoes, cassava and other staple food on their 5 per cent land. Other peasants tried to cultivate land that had been abandoned by the collective.

---

20   Kết quả và kinh nghiệm phát huy quyền làm chủ tập thể ở HTX sản xuất nông nghiệp 1 Điện Nam [Result of and experiences from facilitating collective mastery in Điện Nam Collective No. 1], *Quảng Nam-Đà Nẵng*, 27 June 1979, p. 3.
21   Author's interview, 19 October 2005, Hiền Lộc.
22   Author's interview, 15 October 2005, Hiền Lộc.
23   Some regulations on establishing collectives, *Quảng Nam-Đà Nẵng*, 26 August 1978, p. 1.

Peasants were supposed to harmonise their family economy with the collective economy; however, peasants tended to favour the former because they could see the direct connection between their efforts and the rewards. They therefore devoted as much of their time and resources as possible to their family economy to supplement the food the collective fell short of supplying. Villagers in Hiền Lộc and Thanh Yên recalled that, despite limited individual land, the family economy contributed a great part of their livelihoods. A man in Thanh Yên village recalled:

> When joining the collective, I retained my garden land as 5 per cent land. [The collective granted 5 per cent land to those who did not have enough garden land.] The land was a great help. During the period of the work-points system, our family received only 90–100 kilograms per season from the collective. This amount was enough for my family to consume within one month. But thanks to our 5 per cent land, we grew sweet potatoes and cassavas, which enabled us to survive.[24]

Villagers also commented that people made use of any available resources—for example, time, land and inputs—and invested them in their individual farming. For example, they used quality manure for their own sweet potatoes and cassava plots, while they gave the collective poor manure in exchange for work-points.[25] Some did their collective work fast and carelessly to have more time to devote to their own work. Some made use of land the collective did not use—for example, reclaiming the uncultivated corners of collective land, lake edges, the banks of streams and forest. As one elderly woman told me: 'At that time, we reclaimed land anywhere; we reclaimed even a small piece of land to plant sweet potatoes and cassava.'[26]

Because peasants were concerned with their own interests, there were conflicts between collective and family work. For example, the *Quảng Nam-Đà Nẵng* newspaper reported in January 1981 that, in Quế Sơn collective in Quế Sơn district, 'after transplanting seedlings, the collective leaders were not able to mobilise peasants to weed because they were busy growing cassava in their own gardens'. In response, the collective leaders had to rely on local authorities and mass organisations

---

24 Author's interview, 9 November 2005, Thanh Yên.
25 Author's interview, 21 October 2005, Hiền Lộc.
26 Author's interview, 14 October 2005, Hiền Lộc.

to force people to work.[27] *Quảng Nam-Đà Nẵng* reported in February 1981 that instances of peasants practising 'neglect of common property' had become prevalent in collectives. This led to a situation in which individual plots were lush, while crops in collective plots were stunted and full of weeds.[28]

Villagers recalled that people made use of collective resources for their own family economy. For example, when fertilising collective fields, people often hid some in bushes and took it home later for their own plots.[29] Similarly, when harvesting, carrying, threshing and drying grain, peasants often took some for themselves. Villagers in Thanh Yên recalled children following their parents to glean the rice ears they intentionally dropped when harvesting. When carrying sheaves of grain from the fields to the drying sites, some peasants hid sheaves in the bushes and took them home later. Those who brought kettles of water to the harvesters often returned home with kettles full of grain. When threshing, peasants tried to leave some grain on the stalk so they could thresh it again at home.[30] A brigade leader in Bình Lãnh collective recalled:

> Whenever we did not pay enough attention to watching collective grain, peasants stole it. So, at the harvest time, we had to watch day and night. When harvesting, if checkers were absent, people hid grain in the fields. When threshing, if the checkers were negligent, people often hid the grain in the straw they carried home.[31]

As early as April 1979, *Quảng Nam-Đà Nẵng* reported that 'the phenomena of [peasants] stealing grain and collective property were widespread'.[32] Another article reported:

> When harvesting, there were too many rice-gleaners. Those who carried grain to the drying sites of the brigades often dropped by [to hide grain] in collective members' [relatives' or friends'] houses. When threshing

---

27 Hợp tác xã Quế Tân I: Xây dựng con người, xây dựng hợp tác xã [Quế Tân Collective No. 1: Training people and building the collective], *Quảng Nam-Đà Nẵng*, 21 January 1981, p. 2.

28 Nhìn vào đồng ruộng tập thể: Giống lúa [Looking at collective fields: Rice seeds], *Quảng Nam-Đà Nẵng*, 28 February 1981, p. 2.

29 Chống hao hụt mất mát sản phẩm nông nghiệp khi thu hoạch [Preventing loss of collective produce during harvesting], *Quảng Nam-Đà Nẵng*, 1 December 1981, p. 2.

30 Author's interview, 9 November 2005, Thanh Yên.

31 Author's interview, 19 October 2005, Hiền Lộc.

32 Thành lập 32 hợp tác xã trong vụ Hè-Thu toàn tỉnh có 164 hợp tác xã' [With 32 more collectives established, the province has 164 collectives for the summer–autumn crops], *Quảng Nam-Đà Nẵng*, 19 April 1979, p. 1.

at the brigade's yards, collective members threshed deceitfully and let straw still retain many grains so that after taking the straw home, they could get more grain from it.[33]

Villagers also said the economic efficiency of secondary crops such as peanuts, sugarcane and sweet potatoes was even worse than the rice crop because these plants were often stolen at planting and harvest times. A collective leader in Bình Lãnh recalled that, when sowing peanut seeds, peasants planted the flatter ones and put the full-sized ones into their pockets. When harvesting the peanuts, they ate some and hid some, which significantly reduced the quantity of the harvest.[34] A woman in Hiền Lộc recalled:

> For the peanut crop, the collective leaders did not allow young people to harvest because they feared they would eat too many peanuts. Instead, they used elderly people who were toothless and could not eat much. But they could not keep people from stealing. How can we catch a thief living in our own house? It didn't make sense to keep watching people all the time. They certainly needed to absent themselves. Likewise, when harvesting cassava and sweet potatoes, peasants often hid good ones in the soil and returned to get them later.[35]

In short, despite peasants being labelled 'the masters of the collectives', everyday politics undermined party leaders' expectations. To secure their livelihoods and survive, peasants deployed strategies such as optimising work-points and stealing the collective's resources and produce. The main objectives of these practices were to minimise the disadvantages of collectivisation and maximise their livelihoods. In other words, individually, these actions were merely strategies for peasants' livelihood and survival. However, the aggregate of these individual actions had a powerful political effect because they effectively derailed collectivisation.[36] I will discuss this in more detail in the next sections.

---

33   Preventing loss of collective produce during harvesting, *Quảng Nam-Đà Nẵng*, 1 December 1981, p. 2.
34   Author's interview, 20 October 2004, Bình Lãnh.
35   Author's interview, 15 October 2005, Hiền Lộc.
36   Kerkvliet, *The Power of Everyday Politics*, p. 23.

Local cadres' practices

Despite many cadres being loyal to the VCP's agrarian policies, some in the Central Coast abused their power at the expense of the state's interests. The *Quảng Nam-Đà Nẵng* newspaper reported in May 1979:

> Some party members were bad learners. Some were opportunistic, corrupt, conservative, and small minded, embezzling and colluding. Some displayed bureaucratic, autocratic, and patriarchal behaviours. They made decisions without consulting the masses.[37]

In June 1979, *Quảng Nam-Đà Nẵng* censured its readers:

> Because of inadequate awareness of bad thoughts, some party members were corrupt and self-interested; they were not good examples for the masses. Cadres embezzling [collective property] either individually or collectively were prevalent. Many cadres and party members behaved excessively bureaucratically and were autocratic and aloof … they falsified the actual crop productivity, underreported the output [to the state], poorly managed, stole produce and minimised food contributions to the state [*tính thiệt hơn với nhà nước*].[38]

At the brigade level, some leaders took advantage of their power over the management of labour, costs, production and produce, assigning tasks and giving out work-points to benefit themselves. Villagers in Hiền Lộc and Thanh Yên thought of brigade leaders as 'landlords' who had 'power over life and death' (*quyền sinh sát*). A man in Hiền Lộc village commented:

> The brigade leader was prejudiced [*thành kiến*]. If he disliked someone, he assigned him difficult work. He also took revenge on those who dared to criticise him in public meetings.[39]

In an October 1979 article, *Quảng Nam-Đà Nẵng* reported a typical case of a brigade leader abusing his power in assigning work and giving out work-points. He was accused of stealing the brigade's inputs, 'prolonging work and inflating work-points':

---

37 Để đưa phong trào hợp tác xã nông nghiệp tiến lên mạnh mẽ và vững chắc [To speed up collectivisation forcefully and firmly], *Quảng Nam-Đà Nẵng*, 12 May 1979, p. 1.
38 Tăng cường công tác xây dựng đảng trong các hợp tác xã nông nghiệp [Intensifying building party organisation in the collectives], *Quảng Nam-Đà Nẵng*, 6 June 1979, p. 1.
39 Author's interview, 19 October 2005, Hiền Lộc.

> Regardless of stipulated work norms and work contracts, he gave work-points to collective members at his discretion … if someone gave him a cup of wine, he could increase their tally by 10–20 work-points. He assigned tasks with many work-points to those who were close to him. He also granted five–per cent land to collective members at his discretion. Therefore, many collective members said: 'the brigade leader comes first and the king of heaven second' [*Nhất đội, nhì Trời*].[40]

Brigade leaders were also in charge of collective produce after harvest, so they had more opportunities than others to pilfer some of the produce. A woman in Thanh Yên village claimed that brigade leaders took a considerable amount of collective produce because it was concentrated in their hands.[41] An elderly man said 'some brigade leaders took as much as they liked. They had a party eating chicken and ducks [very valuable food] every night. The people knew, but did nothing.'[42] A former cadre from Bình Định collective no. 2 admitted he colluded with brigade leaders to share the benefits during harvesting time. For example, they underreported the real crop so they could take the difference for themselves.[43]

Villagers in Hiển Lộc and Thanh Yên also complained that collective leaders were in their positions because of their revolutionary credentials, not their education or management skills (*không có trình độ, hồng hơn chuyên*). They did not know how to manage the collective well, and many were self-interested and corrupt. This led to the leakage (*thất thoát*) of considerable amounts of collective property. Some villagers claimed such leakage was greater at the collective level than at the brigade level.

A former brigade leader in Thanh Yên said:

> Leakage was greatest at the collective level. The collective took 60 per cent of the brigade's output and left 40 per cent for peasants. For example, if the brigade harvested 20 tonnes of paddy, the collective took 12 tonnes and left 8 tonnes to distribute among peasants. Therefore, peasants received too little paddy, so they had to supplement their livelihood with growing sweet potatoes and cassava on their

---

40   Xã viên làm chủ phát triển một đối tượng phá hoại hợp tác xã nông nghiệp [Members discover a pilferer in a collective], *Quảng Nam-Đà Nẵng*, 6 October 1979, p. 2.
41   Author's interview, 31 October 2005, Thanh Yên.
42   Ibid.
43   Author's interview, 9 November 2005, Bình Định.

gardens. The collective leaders were supposed to use the produce to buy machines, tractors, fertilisers and to build infrastructure. But they embezzled a great deal through buying these things. For example, when buying a threshing machine, they could embezzle a half of the value by colluding with sellers to write a fake receipt that doubled the actual price. They embezzled 'legally', so the people could not sue them. People saw collective property leaking, so they became disappointed and did not want to work anymore. But they had to work because if they did not, they did not have food to eat.[44]

Quảng Nam-Đà Nẵng reported in November 1979 that collectives in Thăng Bình had incorrectly recorded income, expenditure, inputs and outputs. For example, Bình Nguyên no. 2 and Bình Đào collectives had falsified all records of funds, inventories, cash, receipts and expenditure.[45]

Thanh Yên and Hiền Lộc villagers attributed their low income to the poor quality and great number of collective cadres. They said that, on average, each collective had to feed hundreds of cadres. The salary of each cadre was about 200–300 kilograms of paddy per season, which was much higher than the annual income of the average collective member. Apart from the salary, cadres enjoyed many other benefits, such as attending parties and meetings and buying paddy at low prices (*mua lúa điều hoà*).[46] At that time, the collective spent too much on buying machines and building infrastructure. It subsidised all mass organisations, such as women's unions, peasant associations, schools, hospitals, irrigation teams, specialised and industrial teams, 'priority' families such as those of war martyrs and wounded soldiers, poor peasants and party cells. Even higher-level cadres came to ask for subsidies from the collective.[47]

---

44 Author's interview, 5 November 2005, Thanh Yên.
45 Huyện Thăng Bình tổng kết 2 năm cải tạo nong nghiệp [Thăng Bình district summing up 2 years of agricultural transformation], *Quảng Nam-Đà Nẵng*, 7 November 1979, p. 2.
46 Author's interview, 5 December 2005, Thanh Yên. According to a former collective cadre in Bình Lãnh, the salary of the collective chairman was 140 per cent of the income of the advanced labourer in the collective. The salaries of both the vice-chairman and the chief accountant were equal to 95 per cent of the chairman's salary; the salary of other collective cadres was 90 per cent of the chairman's (Author's interview, 21 October 2004, Bình Lãnh).
47 Author's interview, 9 December 2005, Thanh Yên.

A man in Hiền Lộc commented:

> The collective produced a great deal of produce but 'leakage' was high. Much of the produce was taken to feed a large number of cadres. So, the people often complained: 'the worn rain-hat [peasants] worked so that the pith helmet [cadres] enjoyed' [cởi làm cho cối ăn].[48]

An Giang in the Mekong Delta

Peasants' everyday politics in An Giang's production units

Joining production units, but not participating in work

Unlike many peasants in QN-ĐN, who tended to devote much of their time to collective work to get work-points, many peasants in production units in Long Điền B commune of Chợ Mới district in An Giang were uninterested (thờ ơ) in work-points. Many joined the production units but did not do much collective work. Villagers in Long Điền B recalled that few production unit members were devoted to collective work 'full-time'; most of the full-time workers were poor and landless peasants. Meanwhile, a large number of better-off peasants refused to work or worked only occasionally for production units because they could make a living by doing jobs outside the units or live from their own wealth. A former team leader of production unit no. 1 in the commune recalled:

> Some people joined the production unit simply as a formality [vào hình thức]. They signed up to join but did not go to do collective work, so, at the end of the season, they did not have any work-points to receive paddy. Some families let one or two subsidiary members participate in the production units while the others worked outside [the unit], such as working for wages, fishing or farming elsewhere].[49]

To persuade peasants to do collective work, production units in Long Điền B did not grant household plots (5 per cent land) to peasants as officially stipulated; however, this policy did not help persuade peasants to undertake collective farming. At first, some landless peasants

---

48 Author's interview, 14 October 2005, Hiền Lộc.
49 Author's interview, 27 June 2005, Long Điền B.

were eager to work collectively, but they were later disappointed and dissatisfied with the low rewards and the methods of distribution in the production units. A young formerly landless man recalled:

> At first we worked enthusiastically, but later we felt discouraged. In fact, the production unit produced a considerable amount of paddy but production unit cadres took much of it. Therefore, we received almost nothing. My wife and I were both full-time labourers but the income we received from work-points was not enough for us to survive [*không đủ sống*]. If we worked for wages, we received cash immediately on a daily basis. But for the production unit, we only received paddy at the end of the season. How could the poor live on this? Therefore, some people felt so discouraged that they quit and laboured elsewhere.[50]

A man whose family had five labourers recalled that the income from collective farming was so small his family received only 4–5 *gia* (80–100 kg) of paddy for a whole season. Therefore, he decided to pull out of the unit and made a living elsewhere. He added that many other households had done the same. Because collective farming did not supply adequate food, many people had to take extra jobs outside the unit. Production unit cadres often neglected these peasant practices because they were not able to secure the peasants' livelihoods with collective farming.[51] Similarly, a landless widow with four young children explained why she worked for a production unit for just one month before quitting:

> After reunification, Mr T [the hamlet chief] granted me 4 *công* of land to make a living. Later, at his suggestion, I put all the land into the production unit. I followed others working in the production unit for almost a month, but I did not receive any cash or paddy. My children at home were hungry, so I had to give up doing collective work and laboured for others to raise my little children.[52]

Asked why the rate of peasant participation in collective work was low, many villagers claimed that working collectively was unprofitable and lacked flexibility compared with working individually or for wages. Many better-off families were coerced to join farming collectives, but were

---

50 Author's interview, 17 June 2005, Long Điền B. He meant that the value of a collective farming workday was less than that of a day of wage labour. Local people mentioned that before reunification the value of a day working for wages was about 2 *gia* (40 kg) of paddy. Meanwhile, the value of a collective farming workday was less than 10 kilograms, and even as low as 2–3 kilograms.
51 Author's interview, 28 June 2005, Long Điền B.
52 Ibid.

disappointed (*chán nản*) because they lost their land and the income earned from collective farming was small compared with their previous earnings. They had previously owned considerable amounts of land and enjoyed a better life. A former chairman of production unit no. 2 recalled that his unit actually coerced people into joining, but they did not trust collective farming. Some joined but did not work collectively at all and did not receive any work-points for the whole season; some worked for a few days and earned 10–20 points to avoid being labelled anti-government by the authorities. Some worked for production units for just one or two seasons and then were so disappointed they found jobs elsewhere. So, the percentage of peasants who did collective work in the fields was low—about 10 to 20 per cent.[53] A full-time member of a production unit whose husband was a production unit cadre recalled:

> Others worked only three out of 10 days. We worked 10 out of 10 days. Some better-off people joined but rarely went to work. The production unit had more than 400 *công* [40 hectares] of land and 100 labourers but only about seven to eight people went to do collective work in the field daily. Therefore, we had to work a lot, working to death [*làm muốn chết*].[54]

'The outside foot was longer than the inside foot'

It was common for peasants in the Mekong Delta to try to evade collective farming and make a living outside the production unit. Even some production unit cadres were focused on jobs outside the unit. Those who had boats used them to do trading; those who had relatives in places where collective farming had not yet been established borrowed land to make a living there. Because many places had not yet been collectivised, peasants could easily borrow land to avoid joining collective farming.[55] A former chairman of production unit no. 1 fits this pattern:

> Our production unit was established in 1979. In the first season, the value of work-points was really good [more than 10 kilograms per workday], but after that the value of work-points deteriorated. At the end of 1980, because of the flood, the value of work-points was only 0.7 kilogram. At that time, many peasants left the production unit to

---

53   Author's interview, 30 June 2005, Long Điền B
54   Author's interview, 27 June 2005, Long Điền B.
55   Author's interview, 30 June 2005, Long Điền B.

do outside jobs. But in the following season many of them came back to the production unit because of the increased value of work-points. The higher the value of work-points peasants received, the larger the number of peasants who participated in the production unit.[56]

He also admitted that many peasants and cadres, and even his own family, had to 'keep one foot within and another outside the production unit' (*giữ chân ngoài chân trong*).

This phenomenon was widespread across many collectives in An Giang. The *An Giang* newspaper reported in August 1980 that, when authorities in Mỹ Lương commune in Chợ Mới district established production unit no. 2, many peasants resisted fiercely; some joined but 'kept one foot within and another outside', and the 'outside foot was longer than the inside one' (*chân ngoài dài hơn chân trong*). The number of labourers doing collective work in the fields was very low—sometimes only 20–30 labourers (out of 113) worked in the fields.[57] Similarly, in production unit no. 1 in Chau Long 4 hamlet, some peasants did not do collective farming properly; they also kept 'the outside foot longer than the inside one'. The rate of peasant participation depended on the performance of the production unit. If the unit's performance was good, many engaged in collective work; but, if it was bad, many would leave and find jobs elsewhere. In some families, only one member worked for the production unit, while the remaining members worked outside it to ensure their livelihoods (*xoay sở cuộc sống*).[58]

In short, peasants in Long Điền B and An Giang did not devote the majority of their time, energy and resources to collective farming. Because they lacked confidence in collective farming, many peasants kept 'one foot within and another outside the production unit'. This shows that peasants in An Giang had more options than their counterparts in QN-ĐN to avoid or minimise the disadvantages of collective farming.

56 Author's interview, 28 June 2005, Long Điền B.
57 Về thăm tập đoàn số 2 Mỹ Lương [A visit to Production Unit No. 2 in Mỹ Lương], *An Giang*, 7 December 1980, p. 2.
58 Tập đoàn sản xuất I, khóm Châu Long 4 vững bước tiến lên [Production Unit No. 1, Chau Long 4 Subcommune is progressing], *An Giang*, 9 August 1981, p. 2.

Careless work and neglect of common property

In addition to low levels of participation in collective work, another common problem in Long Điền B production units was the manner in which the peasants worked. They were unenthusiastic and sluggish (*làm không nhiệt tình, làm lê thê*); some just went through the motions, waiting for day's end rather than trying to finish their work. They did not want to work as hard or as carefully as they worked in their own plots.[59] A better-off man in Long Điền B described the manner of peasants' collective farming in his production unit:

> Collective farming according to work-points was poor. People just went through the motions of working without taking care of collective property. When passing by the collective rice plots, if they saw weeds they would not stop and pull them up, as they would have done for their own plots. They worked with their minds elsewhere. People only worked carefully if they worked for themselves. How could the production unit be profitable? I felt sad that our land was pooled for others to work together. But because they were landless and the land was not theirs, they did not love the land at all; they worked for points, so they did not take care of the land. Working together was certainly impossible. I think that only those like Uncle Hồ and Uncle Tôn Đức Thắng could work collectively, but peasants could not. The central leaders were kind; they thought peasants were like them, but peasants were not; they were selfish and different.[60]

Looking back on how peasants resisted collective farming in production units under the work-points system, the *An Giang* newspaper reported, in April 1982:

> When preparing rice seeds to sow, nobody cared whether they were too dry or too soaked. When transporting seeds to the fields, people carried the sacks carelessly and dropped many along the road. When the seeds reached their destination, people did not have enough baskets to take them to the fields. Moreover, people just went through the motions of working until the end of day. When the seeds were ready to sow, they were left sitting in the fields. When it rained slightly, the people refused to work. When it was a bit sunny, many people complained of headaches. After weeding for a while, many people grumbled about backache.[61]

---

59 Author's interview, 28 June 2005, Long Điền B.
60 Ibid. Uncle Hồ is Hồ Chí Minh, the first President of the Democratic Republic of Vietnam; Tôn Đức Thắng, born in An Giang, was Hồ Chí Minh's successor.
61 Vụ lúa khoán ở tập đoàn 3 Tây Khánh B [The results of contracted rice crops in Production Unit No. 3 in Tây Khánh B Commune], *An Giang*, 18 April 1982, p. 3.

The fact that peasants refused to work or worked unenthusiastically affected the operation of production units. People in Long Điền B recalled that the units were not able to mobilise peasants to complete tasks on time, so some fields were left uncultivated and rice plots remained unweeded. A former chairman of Long Điền B production unit no. 1 admitted that his unit was unable to complete weeding in time because of low levels of peasant participation in work schedules: 'weeds were often more numerous than rice shoots' (*cỏ thường nhiều hơn lúa*).[62] A former chairman of production unit no. 2 said:

> For individual farming, peasants prepared the soil and weeded carefully, so the fields hardly had any weeds. But under collective farming, the rice fields were full of weeds because of carelessness. If weeds were not pulled out properly, they would flourish.[63]

Villagers admitted that the collective rice fields were so overgrown with weeds that they looked like a wilderness during the period of collective farming. People in Long Phú hamlet had a popular saying to describe the situation: 'Please come to Long Phú and see weeds that touch the sky' (*Ai về Long Phú mà xem, âm u cỏ rác phủ xanh rợp trời*).[64] A former cadre of Chợ Mới district observed:

> Because peasants did not see efficiency in collective farming, they did not want to work for production units. They were better off leaving the units and finding jobs elsewhere. Therefore, wherever production units existed, the weeds thrived [*tập đoàn đi tới đâu thì cỏ đi tới đó*]. At that time, Mr Do Vuong, a northern cadre, criticised us for not allowing peasants to join production units voluntarily. But I argued with him that no matter how much we propagandised and educated the peasants, they never volunteered to join, because they considered collective farming as working for the sake of cadres.[65]

As well as doing their collective work sloppily and slowly, peasants in Long Điền B were accused by local officials of not caring for and even sabotaging collective property. A former production team leader recalled that floods in 1980 affected the rice fields, so cadres called on

---

62 Author's interview, 28 June 2005, Long Điền B.
63 Author's interview, 30 June 2005, Long Điền B.
64 Author's interview, 27 June 2005, Long Điền B. Long Phú was a hamlet in Long Điền B, Chợ Mới district, An Giang.
65 Author's interview, 22 August 2005, Long Điền B. Châu Thành was one of the districts in An Giang with a low population density. Peasants here had greater economic options to evade collective farming, so the performance of collective farming here was much worse than in other parts of An Giang, such as Chợ Mới district.

people to harvest the crops as soon as possible. No-one responded. People said the rice was not theirs; it belonged to the production unit. He added:

> People were so negative that they even ate sugarcane seedlings [during transplanting] and said anyone who did not eat them was stupid. That was annoying because the production unit had to buy those seedlings. Moreover, when people worked in the fields, they saw broken paddy walls; they should have fixed them. But they did not. They said: 'Why should we when it was not ours?' I ask you, how could the rice survive? A few people had a good attitude but those who had a bad attitude were numerous. Working collectively was certainly impossible.[66]

Similarly, a woman in the production unit recalled: 'We tried to plant sugarcane and corn, but when the crops were ready to harvest, people snitched or destroyed them all.' The production unit then gave up planting secondary crops.[67] A former chairman of production unit no. 2 admitted that secondary crops were a financial failure because of careless cultivation and peasant sabotage. Therefore, in 1980, he decided to give the secondary-crop land to peasants to cultivate individually— similar to what happened in QN-ĐN. In return, peasants paid tax to the state via the production unit.[68]

In summary, unlike their counterparts in QN-ĐN, who tried to strike a compromise with collective farming and pursue work-points, peasants in Long Điền B and An Giang tended to evade collective farming altogether. Some joined the production units but worked infrequently; some worked sluggishly and unenthusiastically and did not care about collective property. These practices significantly affected the performance of collective farming, which will be discussed in the next section.

Local cadres' practices in An Giang

Party leaders accused local cadres in the Mekong Delta of being unenthusiastic about agricultural transformation, having 'weak, messy and slack management of labour, finance, production and distribution of produce' and committing embezzlement. All these factors made

---

66 Author's interview, 27 June 2005, Long Điền B.
67 Author's interview, 28 June 2005, Long Điền B.
68 Author's interview, 30 June 2005, Long Điền B. He argued that his production unit granted secondary-crop land to peasants without the consent of higher authorities. Each household received about 0.5 công (500 sq m) of land to farm individually.

collectivisation in the region difficult.[69] Provincial authorities accused local cadres in An Giang of displaying 'negative practices such as stealing collective property, materials, cash and peasants' work-points, appropriating illegally peasants' land and belongings and bullying the masses' (ức hiếp quần chúng).[70]

Peasants in Long Điền B complained that production unit cadres behaved badly. One man in the commune recalled:

> At that time, cadres enjoyed a comfortable life. They controlled everything such as work-points, materials, cash and paddy; the unit members did not know anything about those things. After harvesting, they controlled all paddy and only distributed part of it to each person according to the amount of work-points. We did not know exactly how they used the remaining.[71]

Some argued that cadres did not make public (công khai) the production unit's income and expenses. They only released one financial report a year, and this was often a 'ghost' (fake) report (báo cáo ma).[72]

The An Giang newspaper describes many cases of embezzlement in production units. For example, 27 inspections in May 1980 found 40 cadres had been embezzling collective property. Authorities received 361 complaints from peasants—most about cadres stealing and bullying.[73] In 1981, Chợ Mới district inspectors discovered that the managerial board of production units in Long Điền B had embezzled collective property. As a result, some production unit cadres were sentenced to a few months' imprisonment.[74] Informants complained that, although some cadres were sacked or imprisoned, authorities were not able to entirely eliminate corruption among the cadres. New cadres might be better behaved initially, but eventually, they committed the same wrongdoings. Some attributed the cadres' problems to policy mechanisms (do cơ chế chính sách) that gave considerable power to cadres in terms of controlling and managing production units.[75]

---

69   BCTNNMN, Report of Central Committee for Agricultural Transformation, p. 4.
70   Tăng cường chỉ đạo công tác chống tiêu cực [Intensifying the fight against negativism], An Giang, 8 June 1980, p. 2.
71   Author's interview, 29 June 2005, Long Điền B.
72   Author's interview, 20 June 2005, Long Điền B.
73   An Giang đẩy mạnh công tác chống tiêu cực [An Giang speeds up the fight against negativism], An Giang, 8 June 1980, p. 2.
74   Author's interview, 28 June 2005, Long Điền B.
75   Author's interviews, June–August 2005, Long Điền B.

A poor man who at first supported the new authorities and worked enthusiastically for the production unit shared his story:

> At that time, the authorities told us that, from now on, people had to join production units to work collectively because individual farming was not allowed. We obeyed and joined to work for the production unit. But the authorities cheated people [*chính quyền lừa dối dân*]. We conformed to the policy while many production unit cadres, even higher officials, left to work individually ... Most cadres were self-interested; they stole collective property with no conscience pangs. It was common that production unit cadres stole collective paddy and were caught by members. As far as I remember, Mr Ba Trực at the Hạt Giang School of Agricultural Transformation said that, if a production unit operated according to socialist principles, it was a heaven on earth for poor households. But if it went wrong, it was much worse and crueller than [life under] previous landlords. He explained that the landlords forced peasants to fill their storehouses full of paddy but people could borrow it back when they needed some. Meanwhile, production unit cadres only focused on [*chỉ có biết*] stealing, pilfering and embezzling collective property without caring about their members. All of these certainly made collective farming go to ruin and peasants suffer starvation.[76]

Some Long Điền B peasants also complained that cadres showed favouritism when assigning work tasks and grading work-points. A poor man whose family joined a production unit but infrequently went to work said production unit cadres showed favouritism to their relatives and friends. These people often received many points because the unit assigned them light tasks worth numerous points and those who were not close to the cadres got less work-points, even though they worked harder. Many therefore wanted to quit and rely on outside jobs to make a living.[77] In other production units, cadres gave the same points to everyone, undermining any incentive people might have to work well. A team leader of production unit no. 1 recalled:

> At first, I was a production unit member. Because I worked hard, I was elected team leader in charge of grading points for the whole team. It was impossible to follow the grading regulations because I feared hurting others' feelings [*sợ mất lòng*]. For example, according to the regulation, if someone came to work one hour late, I had to subtract his

---

76   Author's interview, 30 June 2005, Long Điền B.
77   Author's interview, 29 June 2005, Long Điền B.

work-points. The regulation said so, but in practice, we were afraid of hurting others' feelings so we distributed work-points to people evenly [cào bằng]. At first, some peasants worked enthusiastically but later lost their incentive because there was no difference between those who worked hard and those who worked sloppily.[78]

Local cadres too often mismanaged state resources and did not serve the people responsibly. During the work-points period, the *An Giang* newspaper reported numerous cases of problem cadres. For example, a November 1980 article accused cadres of 'snitching' (*ăn xén*) fertiliser from bags sold to peasants in a state trading shop in Châu Phú district. A bag of fertiliser should have weighed 50 kilograms, but in this district many bags weighed only 46 or 47 kilograms. Peasants also discovered salt and other ingredients had been mixed with the fertiliser.[79]

Figure 5.1 Rice production unit

A worm says to his wife, 'Do not be afraid of moving here, we'll be safe because the production unit manager has already sold all of the pesticides on the black market!'

Source: Drawn by Văn Thành, published in *An Giang*, 22 March 1981.

---

78 Author's interview, 27 June 2005, Long Điển B.
79 Chuyện to nhỏ: Ăn xén của dân [Pilfering people's resources], *An Giang*, 23 November 1980, p. 3.

Similarly, in the winter–spring of 1980–81, cadres in charge of storehouses in Thoại Sơn district embezzled 310 tonnes of paddy, which they simply reported as missing. Cadres colluded with private merchants buying paddy so that both gained financial benefits at the expense of the state and food supplies.[80]

Figure 5.2 Food procurement station

At the food procurement station, a man who sells rice bribes the official so he will ignore the water and sand mixed into his rice. He ponders: 'In life, sometimes a word can increase the weight!'

Source: Drawn by Văn Thành, published in *An Giang*, 3 May 1980.

In explaining the increased prices of paddy in An Giang, a local newspaper reported that some of the cadres responsible for controlling free markets and extending socialist markets were actually corrupt and colluded with private rice merchants, creating favourable conditions for an illicit rice trade. At the same time, cadres 'blocked transport and prohibited markets' (*ngăn sông cấm chợ*) for ordinary labourers.[81] A man in Long Điền B recalled that he went to harvest rice crops for wages (*cắt lúa mướn*) in Thoại Sơn district and took home a few *giạ* of paddy,

---

80 Vài nét về những kho chứa lúa ở Thoại Sơn [Some problems with rice stores in Thoại Sơn], *An Giang*, 23 August 1981, p. 3.
81 Vì sao giá lúa leo thang [Why rice prices escalate], *An Giang*, 27 October 1980, p. 2.

but cadres from a food purchasing station stopped him and told him to surrender his paddy to them. Meanwhile, rice merchants who colluded with the cadres passed through easily.[82] Another man commented:

> Policies said that peasants were not allowed to cultivate and transfer paddy across borders. But if you had money to bribe the cadres you could do this without any difficulty.[83]

Long Điền B peasants and the *An Giang* newspaper accused local cadres of misusing common property. They said cadres frequently organised meetings and parties (*nhậu nhẹt*), wasting time and other resources, which made the state's organisations function poorly and significantly affected people's social and economic activities. The following cartoons help us understand these problems (Figures 5.3–5.6).

Figure 5.3 A farmer and a merchant at a food procurement station

A local officer in charge of preventing private trading points his left hand at a farmer who has two chickens and shouts, 'Hand them over!' Meanwhile, in his right hand, he receives a bribe from a merchant with many bags of rice and beans. She says, 'Here are my permission papers to transport goods.'

Source: Drawn by Nhi, published in *An Giang*, 12 October 1980.

---

82 Author's interview, 9 August 2005, Long Điền B.
83 Author's interview, 30 June 2005, Long Điền B.

Figure 5.4 Drinking at work

A farmer comes to a local office at 2 pm, showing a form to an official, and shyly says, 'Sir, please consider my form.' The officer, who is in the middle of a drinking session, shouts at him, 'Don't you see we are busy with our meeting?'

Source: Drawn by Văn Thành, published in *An Giang*, 16 November 1980.

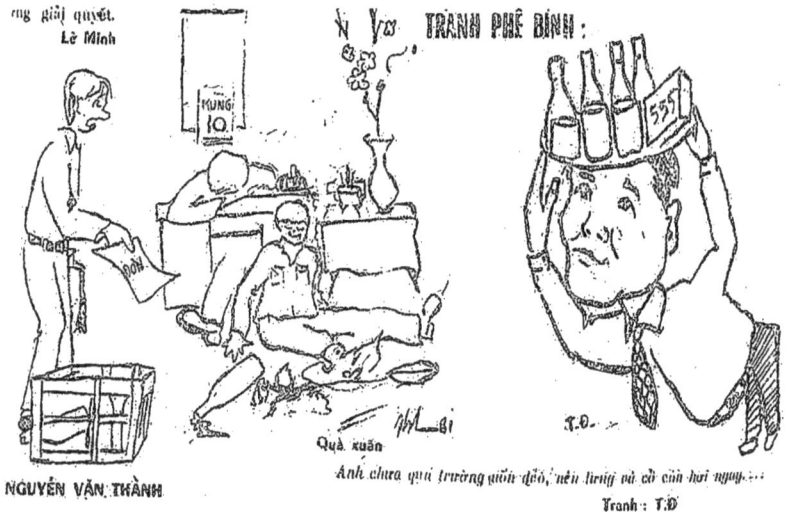

Figure 5.5 Tet (New Year) gifts

An officer submits a form to a higher official on New Year without including a 'gift', and is criticised for not behaving properly (like the man on the right).

Source: Drawn by Văn Thành, published in *An Giang*, 22 February 1981.

Figure 5.6 Smuggling
The ambulance carries a patient surrounded by smuggled MSG, textiles and cigarettes.
Source: Drawn by Văn Thành, published in *An Giang*, 4 January 1981.

The performance of collective organisations under the work-points system

QN-ĐN in the Central Coast region

In the first years after reunification, staple food production in QN-ĐN reportedly increased rapidly, from 149,062 tonnes in 1975 to 380,000 tonnes in 1978. Inspired by this achievement, QN-ĐN officials believed that, under their close leadership, the province could produce 550,000 tonnes of staple food by the end of the 1976–80 five-year plan. They believed that only collectivisation with 'three revolutions' would enable agriculture to meet that target.[84]

---

84   29-3-1975 – 29-3-1979: 4 năm lớn mạnh về kinh tế [From 29 March 1975 to 29 March 1979: 4 years of economic expansion], *Quảng Nam-Đà Nẵng*, 28 March 1979, p. 1.

However, in contrast to their expectations, when collectivisation in the province was extended, food production stagnated and did not match the increases in the area under cultivation and in agricultural investment. A leading article in *Quảng Nam-Đà Nẵng* in September 1979, titled 'Some urgent measures to increase food production', reported: 'The rice productivity of the spring–summer of 1979 is low while the coming summer–autumn is under the threat of drought and flood. Starvation has occurred in some locations.' The article argued that, in addition to bad weather, the poor performance of staple food production was because local authorities mismanaged and underutilised agricultural land (especially secondary-crop land) and labour. To improve food production, the article urged collectives to temporarily lend secondary-crop land to collective members for three years.[85] Similarly, the chairman of the provincial Committee for Agricultural Transformation admitted that the area and yield of secondary crops had decreased compared with pre-collectivisation times, and he urged collectives to lend secondary land to their members.[86] Hiền Lộc and Thanh Yên also recalled secondary crops doing badly because of peasants pilfering produce. Finally, in 1980, collective leaders decided to temporarily redistribute secondary-crop land to households.[87]

Accounts in *Quảng Nam-Đà Nẵng* showed that the province continued to have bad harvests in the winter–spring of 1979–80; average rice productivity was about 2.5 tonnes per hectare compared with 2.92 in the previous winter–spring, of 1978–79. Thousands of hectares of rice yielded no crop (*mất trắng*). For example, Tam Kỳ district suffered failed rice crops across 557 hectares, while Quế Sơn district had 188 hectares producing no harvest.[88] Similarly, of Thăng Bình district's 4,500 hectares of rice in the winter–spring of 1979–80, 800 hectares yielded no crop and 1,600 hectares returned poor yields. The average rice productivity in Thăng Bình district fell to 1.47 tonnes per hectare.[89]

85   Các biện pháp cấp bách đẩy mạnh sản xuất lương thực, thực phẩm ổn định đời sống nhân dân [Some urgent measures to increase food production], *Quảng Nam-Đà Nẵng*, 15 September 1979, p. 1.

86   Nhận thức đúng đắn và thi hành nghiêm chỉnh việc tạm giao đất chuyên trồng màu cho xã viên sản xuất [Understanding well and seriously implementing a temporary redistribution of secondary land to members], *Quảng Nam-Đà Nẵng*, 19 September 1979, p. 1.

87   Author's interviews, October–December 2005, Thanh Yên.

88   Vụ Đông-Xuân 1980–1981 được mùa cả lúa và màu [Good rice and secondary crops harvested in winter–spring of 1980–1981], *Quảng Nam-Đà Nẵng*, 23 May 1981, p. 1.

89   Huyện Thăng Bình phấn đấu đạt 65000 tấn lương thực năm 1981 [Thăng Bình is striving to produce 65,000 tonnes of food], *Quảng Nam-Đà Nẵng*, 9 September 1981, p. 2.

As a result of the efforts of QN-ĐN authorities to expand irrigation and agricultural land, and increase the number of crops per year and the use of chemical fertilisers, by 1980, the province's staple food production was expected to reach 460,000 tonnes of paddy.[90] However, according to a recent report from the Quảng Nam Department of Agriculture and Rural Development (Sở Nông Nghiệp Phát Triển Nông Thôn Quảng Nam, or SNNPTNTQN) QN-ĐN's grain production (including rice and corn) in 1980 was just 285,426 tonnes of paddy equivalent—falling short of the target.[91]

Many other provinces in the Central Coast region faced similar food production shortfalls. In assessing the effects of collectivisation on the Central Coast's agriculture, one study found:

> The Central Coast was the region in which collectivization occurred most quickly and thoroughly and was most like the northern models. In this region, all peasants' means of production became collective property; labor was tightly controlled by centralized leadership; household economy is highly restricted and even prohibited (so, they generated only a little staple food). Therefore, during the peak period [of] collectivization, the region faced a severe problem of staple food production. For example, thousands of hectares of secondary crop land in Thăng Bình district of QN-DN were abandoned in 1978.[92]

Thanh Yên and Hiền Lộc villagers recalled their living conditions dramatically deteriorating during the work-points period. At the beginning of collective farming, the value of a workday in Bình Định collective no. 2 was 0.5 kilograms of paddy; later, it fell to 0.3 kilograms. Bình Lãnh collective faced a similar situation: the value of a workday fell from 3 kilograms of paddy in 1977 to 2 kilograms in 1978, 1.5 kilograms in 1979 and 0.5 kilogram in 1980.[93]

An elderly man in Thanh Yên village recalled that collective farming caused hunger, and the value of a workday—0.3 to 0.5 kilogram of paddy—was not enough to 'feed a rooster', let alone a person.[94] According to a former brigade leader of Bình Định collective no. 2:

---

90 Sở Nông Nghiệp Phát Triển Nông Thôn Quảng Nam (2005), *Kết quả sản xuất nông nghiệp năm 1976–2004 [Agricultural Production 1976–2004]*, Tam Kỳ: Sở Nông Nghiệp và Phát Triển Nông Thôn Quảng Nam. I received this report during fieldwork in Quảng Nam in 2005.
91 CTKQN, *Quảng Nam's Socioeconomic Development*.
92 Nguyễn Sinh Cúc, *Agricultural and Rural Development in Vietnam*, p. 32.
93 Author's interview, 9 December 2005, Thanh Yên.
94 Author's interview, 31 October 2005, Thanh Yên.

The value of a workday during the first harvest of the collective [in the summer–autumn of 1979] was 0.5 kilogram of paddy. In the following season, the winter–spring of 1979–1980, the collective had such a bad harvest that collective cadres had to go elsewhere to buy food for their families. The value of a workday in that season was less than 0.3 kilogram of paddy. In the summer–autumn of 1980 season, the harvest was also bad. In the winter–spring of 1980–1981, the collective enjoyed a good harvest but the district's authorities took a large quantity of collective paddy, so the value of a workday never reached 0.5 kilogram of paddy during the work-points system.[95]

Hiền Lộc and Thanh Yên villagers often resorted to sarcasm to describe their living conditions during the work-points period. For example, I recorded statements such as: 'collective farming produced so little rice that, when eating, people had to lick rice clinging to their chopsticks' (lúa điếm là liếm đũa); 'working for the collective, there were no clothes to cover one's privates' (hợp tác hợp te không có miếng vải mà che cái lỗn); 'in the evening, people had dinner with sweet potatoes to sleep; in the morning, people had breakfast with sweet potatoes to work; at noon, people opened their mouth to chew sweet potatoes again' (tối ăn khoai đi ngủ, sáng ăn củ đi làm, trưa về hả hàm nhai khoai); and 'farmers work and cadres enjoy' (cời làm cho cối ăn).[96]

Villagers recalled people living on sweet potatoes and cassava. An elderly man in Hiền Lộc told me his family did not have enough rice so they had two meals of cassava a day and one of rice mixed with cassava.[97] An elderly man in Thanh Yên shared a similar story:

At that time, we substituted sweet potatoes for rice. Sweet potatoes were our main staple food. During one season of working for the collective, my family received only 20 kilograms of paddy. How could we live? The collective took much of what we produced.[98]

---

95 Author's interviews, 22 and 23 October 2005, Hiền Lộc.
96 Lúa điếm là liếm đũa is a kind of backwards slang: lúa điếm is the amount of rice peasants received according to their work-points; liếm đũa literally means 'licking chopsticks', implying rice was so scarce that, after meals, people were still hungry. Hợp tác hợp te không có miếng vải mà che cái lỗn is a modified version of the government slogan 'working for collectives, few people went on foot; many went by bus' (hợp tác hợp te đi bộ thì ít đi xe thì nhiều).
97 Author's interview, 22 October 2005, Hiền Lộc.
98 Author's interview, 31 October 2005, Thanh Yên.

Villagers believed the poor performance of collective farming (its low productivity), high levels of leakage (*thất thoát*) and the wasting of collective resources (*lãng phí*) were the main reasons for their deteriorating living conditions. According to villagers, collective farming always produced poor harvests because people did not do collective work as carefully as they did private work and merely went through the motions of working; they did not take care of collective fields, and the ploughing, spreading of manure and weeding were done carelessly and unevenly. Some plots produced good harvests while neighbouring ones were bad.[99] A man in Thanh Yên commented:

> If collective farming had continued, land would become unsuitable for ploughing and transplanting any more because the soil would become harder and suffer degradation. Moreover, for years of collective farming, the collective plots would significantly decrease in size because people did not plough the soil properly; they did not hoe the corners and clear the edges properly.[100]

In the opinion of many villagers, the poor productivity of collective farming was largely a result of people's everyday politics and survival strategies. They also considered the 'leakage' from and waste in collectives, caused by cadres, as other major reasons for their low income. The leakage, villagers said, resulted not only from peasants' theft and cadres' embezzlement, but also from extraction by other individuals, mass organisations and the state. They argued that, because a large amount of collective produce was extracted to support cadres, subsidise mass organisations and pay state taxes and obligations, collective members received little income. Despite the fact that members were supposed to share more than 60 per cent of collective produce, leakage and waste meant they received less than half of this.[101] This problem was not confined to Bình Lãnh and Bình Định collectives, but was common in many Central Coast collectives. According to a report by the Committee for Southern Agricultural Transformation in November 1979:

> In some Central Coast locations, the state's share in collective food distribution was about 30 to 40 per cent, together with collective funds and supplies for local guerillas, local cadres, party and mass organisations which meant that collective members received less than 60 per cent of

---

99   Author's interview, 15 October 2005, Hiển Lộc.
100 Author's interviews, 10 October 2004; 5 October 2005, Thanh Yên.
101 Author's interviews, October–December 2005, Thanh Yên and Hiển Lộc.

total produce, as regulated. Even in some locations collective members received only 40–50 per cent of produce. Meanwhile, peasants' secondary crops produced a bad harvest. So, the living standards of collective members were very low; many households faced difficulties in earning enough to live. Starvation occurred in some places such as Tam Ka district, QN-ĐN.[102]

Another problem was poor governance, especially the inefficient management of collective resources. Villagers argued that collectives were too large and cadres could not control resources (such as agricultural inputs, land and labour), production and output. Since a collective was unable to utilise all of the available agricultural land and labour, some land was left uncultivated or was cultivated too late. Moreover, under collective farming, workers were not able to weed, care for fields or dry produce as efficiently as under individual farming. A Hiền Lộc villager recalled delivering grain to the brigade's house and then to the collective's storehouse. Even though the grain had not dried properly, it was still put into storage. Later much of this produce rotted—an example of how collective farming wasted a lot of resources.[103] Another man argued that 'the state thought that centralised leadership and management made agriculture stronger, but it failed to do so. I thought that individual farming was much more efficient than collective farming.'[104]

An Giang in the Mekong Delta

As discussed in the previous section, peasants from Long Điền B often expressed their objections to collective farming by rarely undertaking collective work or doing it unenthusiastically. Combined with local cadres' mismanagement, this led to poor performance for collective farming.

Long Điền B peasants said the work-points system was terribly poor compared with their individual farming methods. The common reasons they gave were that 'people did collective work unenthusiastically and sluggishly', 'no-one took care of common property', 'production unit cadres embezzled collective resources' and 'management of the

102 BCTNNMN, *Report of Central Committee for Agricultural Transformation*, p. 2.
103 Author's interview, 19 October 2005, Hiền Lộc.
104 Author's interview, 20 October 2005, Hiền Lộc.

production units was slack'. As a result, peasants' incomes were even worse than those of tenants in the period of landlords. An elderly man argued:

> In the French time, the tenants who did not have land could farm on the landlord's; the rent was not too much and tenants could make a living. Furthermore, at that time, wild fish were still numerous, which enabled people to make a living easily [dễ sinh sống]. When joining the production unit, people worked miserably and results were low. Because people did not want to work collectively, they did collective work sloppily [làm khơi khơi] and weeded carelessly, so weeds overgrew. Therefore, at that time, peasants' income was less than previously.[105]

A former chairperson of production unit no. 1 in Long Điền B recalled that, initially, a collective workday was valued at more than 10 kilograms, but this later decreased. At the end of 1980, after heavy floods, the value of a workday was only 0.7 kilogram (the same as in QN-ĐN). Peasants were disappointed and wanted to quit collective farming.[106] A landless man who worked full-time for production units recalled that he and his wife received about only 20 giạ of paddy (400 kg) for a whole season— not enough to feed his family and much less than his previous income from wage labour.[107]

The low value of a workday was not limited to collective farming in Long Điền B, but occurred across many parts of An Giang. Production unit no. 3 of Tây Khánh B commune in Long Xuyên faced the same situation. A full-time and hardworking member received only 10 giạ (200 kg) of paddy per season. Here again, the reason for the low income was that people did not participate in collective farming wholeheartedly, but tried to make a living outside the production unit.[108] A Chợ Mới district official who had experience of collectivisation in the period 1979–81 observed:

---

105 Author's interview, 20 June 2005, Long Điền B.
106 Author's interview, 28 June 2005, Long Điền B.
107 Author's interview, 23 June 2005, Long Điền B.
108 The results of contracted rice crops, An Giang, 18 April 1982, p. 2.

The living conditions of peasants in collective farming production units deteriorated. Where local authorities carried out collectivisation exactly according to the state policy, peasants faced many more difficulties in making a living. But where local authorities loosely applied the policy of collectivisation, peasants found it easier to make a living.[109]

An Giang provincial Resolution No. 017/NQ-TU (26 November 1981) admitted that, 'in many production units and collectives, production had not increased, it had even decreased; the living conditions of production unit members have not improved'.[110]

In June 1981, the *An Giang* newspaper reported that rice productivity in the Tây Huế collective in the winter–spring of 1979–80 was about 1.5 tonnes per hectare (compared with the 4–5 tonnes per hectare achieved under individual farming). Moreover, leakage (*thất thoát*) and cadre embezzlement accounted for 50 per cent of this produce. The living conditions of collective members therefore had worsened.[111] Another example was Phú Quý production unit in Phú An commune in Châu Phú district, whose land was assessed as fertile and, before collectivisation, had achieved average rice productivity of about 5–6 tonnes of paddy per hectare. However, after collectivisation, rice productivity fell to 1–2 tonnes of paddy per hectare.[112]

By 1981, collectivisation in An Giang involved less than 10 per cent of agricultural land and peasant households, and the effect of collectivisation policy on the province's agricultural production was minor compared with that in QN-ĐN in the Central Coast. However, the combined effects of agricultural transformation— including collectivisation, prohibition of non-resident cultivators, land redistribution, conversion to double cropping, the low prices paid for grain procurement and free market restrictions—significantly hindered the development of An Giang's agricultural sector. For example, the low prices paid for grain procurement discouraged peasants from increasing production. Instead, they produced only enough grain for their own family's consumption. The prohibition of non-resident cultivators limited the productive capacity of peasants who previously

109 Author's interview, 23 June 2005, Chợ Mới.
110 Võ Tòng Xuân and Chu Hữu Quý, *KX Account 08-11*, p. 33.
111 Chuyển biến mới ở HTX Tây Huế [Good progress in Tây Huế collective], *An Giang*, 7 June 1981, p. 2.
112 Vài nét về một tập đoàn yếu kém [Some portraits of a weak production unit], *An Giang*, 6 September 1981, p. 2.

enjoyed relative freedom in choosing where to live and selecting their own businesses. According to Nguyễn Minh Nhị, a former An Giang party secretary, from 1976 to 1979, food production in the province stagnated at about 500,000 tonnes of paddy equivalent per annum. Due to heavy floods in 1978, An Giang's food production fell to less than 400,000 tonnes and starvation occurred in some places.[113] From 1979 to 1980, the collapse of the majority of production units, which released peasants and land from collective farming, contributed to a slight increase in food production and productivity in 1980 and 1981. In general, despite much effort to modernise, food production and productivity during 1975–81 did not increase as much as expected (see Table 5.1).

Table 5.1 An Giang's cultivated area, food yield and rice productivity, 1975–81

	1975	1976	1977	1978	1979	1980	1981
Area of annual food crops (hectares)	224,572	232,174	254,648	250,402	252,111	303,882	301,099
Food production (tonnes)	465,465	496,286	476,500	363,192	525,814	737,874	691,561
Rice productivity (tonnes per hectare)	2.13	2.25	1.97	1.55	2.27	2.52	2.34

Source: Cục Thống Kê An Giang (CTKAG) (2005), *Tổng hợp diện tích, năng suất sản lượng cây trồng hàng năm và số lượng gia súc gia cầm gia đoạn 1975–2005* [*Area, Productivity and Output of Annual Crops in An Giang from 1975–2005*], Long Xuyên: Cục Thống Kê An Giang.

The stagnation of food production occurred not only in An Giang, but also across the whole Mekong Delta region. According to Nguyễn Sinh Cúc, before 1975, the Mekong Delta was one of the largest commodity rice–producing regions. However, after reunification, in the period 1976–80, despite peaceful times, the region's staple food production did not increase, but fluctuated. In particular, rice production in the region fell between 1976 and 1978 and increased slightly between 1979 and

---

113 Nguyễn Minh Nhị, *An Giang*, p. 1. Some provincial officials explained that, apart from collectivisation and other policies, at that time, cheap prices for food discouraged peasants from increasing their own food production. They produced only enough for their own consumption (Author's interview, 6 June 2005, Long Xuyên).

1980 (Tables 5.2–5.4). Nguyễn Sinh Cúc argued that, apart from bad weather, the fall in rice production was closely linked to the expansion of collectivisation. Moreover, he attributed the slight increase in rice production in 1979–80 largely to the collapse of a large number of production units, which released peasants and considerable amounts of land from collective farming.[114]

In general, food production in the region fell short of the VCP leaders' expectations, and aggravated the severe food shortage across the country in the early 1980s.

Table 5.2 Rice crop area, paddy production and rice productivity in the Mekong Delta, 1976–80

	1976	1977	1978	1979	1980
Area of rice crop (thousand hectares)	2,062	2,099	2,062	2,086	2,096
Paddy production (thousand tonnes)	4,206	3,478	3,565	4,650	4,835
Rice productivity (tonnes per hectare)	2.04	1.66	1.73	2.23	2.3

Sources: Nguyễn Sinh Cúc (1991), *Thực Trạng Nông Nghiệp, Nông Thôn và Nông Dân Việt Nam 1976–1990 [Agricultural and Rural Development in Vietnam 1976–1990]*, Hà Nội: NXB Thống Kê; and Cục Thống Kê An Giang (CTKAG) (2005), *Tổng hợp diện tích, năng suất sản lượng cây trồng hàng năm và số lượng gia súc gia cầm giai đoạn 1975–2005 [Area, Productivity and Output of Annual Crops in An Giang from 1975–2005]*, Long Xuyên: Cục Thống Kê An Giang.

Performance of Vietnam's agriculture in 1976–80 and modification of the national agrarian policy

According to Nguyễn Sinh Cúc, from 1976 to 1980, land redistribution and collectivisation had a negative effect on the south's agriculture. He argued that peasants' negative practices resisting collectivisation (such as abandoning their land and neglecting to care for rice fields), egalitarian land redistribution and cadres' corruption contributed significantly to the poor performance of agriculture in the south. In particular, paddy productivity and yield stagnated during 1976–80 (see Table 5.3).

---

114 Nguyễn Sinh Cúc, *Agricultural and Rural Development in Vietnam*, pp. 31, 32.

Table 5.3 Area, productivity and output of rice crops in southern Vietnam, 1976–80

	1976	1977	1978	1979	1980
Area of rice (thousand hectares)	2,909	3,034	3,010	3,011	3,236
Output (thousand tonnes)	6,346	5,887	5,014	6,431	7,207
Productivity (tonnes per hectare)	1.97	1.94	1.67	2.14	2.23

Source: Nguyễn Sinh Cúc (1991), *Thực Trạng Nông Nghiệp, Nông Thôn và Nông Dân Việt Nam 1976–1990* [*Agricultural and Rural Development in Vietnam 1976–1990*], Hà Nội: NXB Thống Kê, p. 8.

Ben Kerkvliet's study of northern Vietnam showed that, due to everyday politics regarding land, labour and harvesting, staple food production there decreased between 1974 and 1980. In particular, paddy production fell by 20 per cent, while staple food per capita decreased from 276 kilograms in 1974 to 215 kilograms in 1980. Moreover, the performance of collective organisations deteriorated during the period 1976–80 and many could not meet tax and other obligations to state agencies.[115]

In general, Vietnam's agriculture and staple food production stagnated in the period 1976–80. The country could not meet many of the targets in the government's 1976–80 five-year plan. Staple food production, for example, reached only 68.5 per cent of the target. Food production could not meet the needs of consumption and inputs for industry. Vietnam had to increase its food imports from 1.2 million tonnes in 1976 to 2.2 million tonnes in 1979.[116]

Table 5.4 Vietnam's staple food production, 1976–80

	1976	1977	1978	1979	1980
Staple food production (thousand tonnes)	13,400	12,579	12,255	13,986	14,382
Staple food per capita (kilograms per person)	274	250	237	266	267
Paddy production (thousand tonnes)	11,828	10,576	9,789	11,362	11,047
Paddy productivity (tonnes per hectare)	2.23	1.95	1.79	2.07	2.11

Source: Nguyễn Sinh Cúc (1991), *Thực Trạng Nông Nghiệp, Nông Thôn và Nông Dân Việt Nam 1976–1990* [*Agricultural and Rural Development in Vietnam 1976–1990*], Hà Nội: NXB Thống Kê, p. 8.

---

115 Kerkvliet, *The Power of Everyday Politics*, pp. 174–5.
116 Nguyễn Sinh Cúc, *Agricultural and Rural Development in Vietnam*, p. 9.

Vietnam in the late 1970s faced persistent food shortages and widespread hunger that alarmed national leaders.[117] Other aspects of the economy also were in bad shape. The growth rate of gross domestic product in industry was −4.7 per cent in 1979 and −10.3 per cent in 1980. Additionally, from late 1978, Vietnam was at war with Cambodia and, from early 1979, it endured armed conflict with China, which consumed high levels of the country's resources.[118]

From 1979 to 1981, VCP leaders released a series of directives and instructions urging local authorities to strengthen collective farming and crack down on local negativism. However, the formation of collective farming had not improved and local malpractices had not disappeared, but increased over time. For example, in QN-ĐN, despite several campaigns to improve collectives and correct cadres' and peasants' negativism, progress was modest, and some problems were only temporarily corrected. The success of these campaigns was uneven and, in general, lower than expected.[119] Many collectives in QN-ĐN were unable to adopt in full the northern model of collectivisation, despite the best efforts of the authorities. With inadequate capacity and poor governance, collectives tended to allow more room for local practices, policy modification such as 'hidden contracts' (khoán chui) or return of some collective land to peasants. Collectives became sites of perennial conflict between cadres, the state and peasants.

In An Giang, despite the government campaigns, the quality and performance of most production units did not improve as much as provincial leaders had expected. By mid-1981, only 40 per cent of production units and collectives in the province farmed collectively and the remainder farmed individually or under other illegal arrangements.[120] By the end of 1981, An Giang had 357 production units and six collectives, but only 35 of the units were 'advanced'. The lead article of the An Giang newspaper in September 1981 reported that collectivisation in the province had been uneven and was not extensive. The number of strong production units and collectives was

---

117 Ibid., p. 9; Phạm Văn Chiến, History of the Vietnamese Economy, p. 159.
118 Kerkvliet, The Power of Everyday Politics, pp. 176–7; Phạm Văn Chiến, History of the Vietnamese Economy, pp. 155–6.
119 Quyết tâm đưa cuộc đấu tranh chống tiêu cực trong năm 1981 lên thành cao trào quần chúng, đều khắp vững chắc [Be resolute in fighting 'negativism' comprehensively in 1981], Quảng Nam-Đà Nẵng, 14 January 1981, p. 1.
120 Đẩy mạnh công tác cải tạo nông nghiệp [Speeding up agricultural transformation], An Giang, 7 June 1981, p. 1.

small. Some policies—such as those relating to land compensation, the family economy and the non-resident cultivator prohibition—had not been implemented correctly or seriously.[121] Another article reported that, despite the efforts of the central and provincial governments, the 'phenomenon of negativism' (*hiện tượng tiêu cực*) remained severe. Negativism included widespread embezzlement and the theft of collective property, wages and work-points.[122]

In response to the food crisis, the poor performance of collective organisations and widespread local use of 'illegal contracts' (*khoán chui*) in collectives across Vietnam, the VCP released Directive No. 100 in January 1981, calling for the expansion of new farming arrangements called 'the product contract to individual workers or groups of workers' (*khoán sản phẩm đến nhóm và người lao động*). These arrangements largely approved local practices and marked a significant modification of Vietnam's agrarian policies.

Under the new system, each peasant household was allocated several small fields and a quota (*mức khoán*) for how much each field should produce. Frequently the quota was 10–15 per cent more than the average production during the previous three to five years. Collective leaders also determined which phases of farmwork should be done collectively and which individually. Farmwork was often divided into eight major phases, with individuals responsible for the three that most closely affected the end product (*sản phẩm cuối cùng*): planting (*trồng*), tending (*chăm sóc*) and harvesting (*thu hoạch*). The remaining phases—considered 'technically complicated'—were preparing the land, providing seeds, irrigation, fertilisation and preventing and controlling disease. Specialised teams and brigades undertook these tasks collectively.[123] The income of collective members was supposed to come from two main sources: work-points earned in collective work and income from the amount each household produced beyond their quota.[124]

---

121 Xã luận: Công tác cải tạo nông nghiệp tỉnh An Giang [The editorial: Agricultural transformation in An Giang], *An Giang*, 6 September 1981, p. 1.
122 Intensifying the fight against negativism, *An Giang*, 8 June 1980, p. 1.
123 Lê Thanh Nghị, *Improving the Product Contract*, p. 645; Kerkvliet, *The Power of Everyday Politics*, p. 193.
124 BCHTU (1993), Chỉ thị cải tiến công tác khoán, mở rộng khoán sản phẩm đến nhóm lao động và người lao động trong hợp tác xã nông nghiệp (ngày 13 tháng 1 năm 1981) [Directive on improving the contracting of products to labour groups and labourers in agricultural cooperatives (13 January 1981)], in Bộ Nông Nghiệp and Công Nghiệp Thực Phẩm (eds), *Chủ trương chính sách của Đảng, Nhà nước và tiếp tục đổi mới và phát triển nông nghiệp và nông thôn* [*Vietnam's Agrarian Policies*], Hà Nội: NXB Nông Nghiệp; Kerkvliet, *The Power of Everyday Politics*, p. 184.

Conclusion

As in the north, in both QN-ĐN and An Giang provinces, everyday peasant politics during 1977–81 significantly affected the performance of collective farming.[125] It is true that collective organisations in both provinces became sites of struggle between peasants, collectives and the state over production, distribution and the balancing of different interests. Like their counterparts in northern collectives, peasants in QN-ĐN tried to maximise the number of work-points they earned rather than production. Meanwhile, peasants in An Giang tended to evade collective farming and focused more on jobs outside the collective to make a living. Local cadres in both provinces also committed various wrongdoings and often took advantage of their power to benefit themselves rather than collectives and the state. While QN-ĐN cadres seemed to strictly control peasants' economic activities and did not allow them to conduct businesses outside the collective, their counterparts in An Giang seemed to be slack in their management and allowed peasants to 'put one foot inside and the other foot outside' the collective. The latter were, however, engaged in higher levels of embezzlement and other negative practices. For example, they frequently organised drinking sessions (*nhậu nhẹt*), stole collective inputs and resources and wasted time. Despite authorities in both provinces putting great effort into correcting peasants' and cadres' 'negative practices', those behaviours increased over time. As a result, collective farming performed poorly in both QN-ĐN and An Giang and food production fell short of local and VCP leaders' expectations.

At the national level, too, despite several campaigns by the VCP to improve collective organisations across the country, performance fell short of expectations. Faced with a crisis in staple food production and stagnation in the agricultural sector, national leaders finally abandoned the work-points system in early 1981 and approved previously 'illegal' local farming arrangements, called the product contract system. The VCP leadership believed that, by using the economic incentive of producing beyond quotas, the new system would motivate peasants to work enthusiastically and efficiently and increase productivity, thereby strengthening and perfecting collective farming.

---

125 Kerkvliet, *The Power of Everyday Politics*, p. 28.

However, the product contracts merely marked a new phase in Vietnam's agriculture in which collective farming again gradually departed from the VCP's original intentions. This will be discussed in the next chapters.

6

Adopting the product contract system and the continuation of land reform and collectivisation, 1981–88

Introduction

Starting in 1981, product contracts—officially called 'The Product Contract to Groups of Workers and Individual Workers' (*Khoán sản phẩm đến nhóm và người lao động*)—became the backbone of collective organisations in Vietnam.[1] According to Đặng Phong, the product contract directive was issued by the Secretariat of the Central Committee of the Vietnamese Communist Party (VCP) (Ban Bí Thư) rather than the Politburo (Bộ Chính Trị) because there were still some differences about this issue among members of the Politburo. The VCP Secretariat believed the product contract system was not a departure from collective farming, but rather an improvement to it. Under the VCP's Directive No. 100 (issued on 13 January 1981), collective organisations were asked to continue perfecting the existing system by more strictly applying the rules for reward and punishment at the brigade level.[2]

---

1 BCHTU, Directive on improving the contracting of products.
2 Đặng Phong, *The Economics of Vietnam*, p. 224.

At the same time, VCP leaders acknowledged the failure of socialist transformation in the previous five years. Persistent in their task of building socialism in the south, party leaders emphasised the continuation of the push for socialist large-scale production and considered collectivisation and agricultural development top priorities for the next socioeconomic five-year plan, 1981–85.[3] The fifth party congress in 1982 also officially changed the target for completing collectivisation in the Southern Region (which includes the Mekong Delta and the South-East Region) from the end of the 1976–80 five-year plan to the end of the 1981–85 five-year plan.

From 1981, the Southern Region had intensified both collectivisation and land redistribution, which encountered weaker resistance from peasants under the product contract system. However, policy implementation fluctuated periodically and largely depended on the VCP's campaigns and directives. With continuous pressure from the VCP, local authorities in the Southern Region were able to complete basic collectivisation by the mid-1980s, although many collectives fell short of expectations. By February 1985, An Giang province was halfway towards completing collectivisation. To achieve full collectivisation in time to coincide with the tenth anniversary of Vietnam's reunification, provincial leaders pushed even harder and, by the end of May 1985, 80 per cent of agricultural land had been collectivised—a minimum index for basic collectivisation.[4]

This chapter will discuss how local authorities in QN-ĐN and An Giang adopted and extended the product contract system and how peasants and local cadres responded to it. It also examines how and why VCP leaders persistently pressed authorities in the Southern Region to achieve socialist agricultural transformation and how those local authorities, especially in An Giang, coped with obstacles to their efforts to complete the VCP's collectivisation goals.

---

3 Vo Nhan Tri, *Vietnam's Economic Policy Since 1975*, p. 126.
4 Đưa phong trào hợp tác hóa của tỉnh nhà lên vững chắc [Advancing collectivisation firmly], *An Giang*, 7 June 1985, p. 1.

Adopting product contracts

Quảng Nam-Đà Nẵng (QN-ĐN), in the Central Coast region

Soon after Directive No. 100 was issued, QN-ĐN leaders held meetings to discuss product contracts and prepare their own directive to guide local authorities. The directive urged each district in QN-ĐN to select one collective in which to experiment with product contracts on paddy fields. To avoid any deviation, it warned that the product contract policy was not intended to redistribute collective land to be farmed individually nor to make a 'blank or full contract' (*khoán trắng*) allowing peasants to undertake all phases of farming. The directive also outlined five principles with which local authorities had to conform in implementing the policy. First, each collective was required to manage and control the collectivised means of production (land, farm tools, draught animals, fertilisers, and so on); no collective was allowed to return the collectivised means of production to members. Second, the collective was required to manage and monitor labour. Third, the collective had to make a production plan based on the district plan. Fourth, the collective had to control the end product and distribute it in an equitable and appropriate way. Finally, the collective had to facilitate members' collective mastery of management and production.[5]

According to a *Quảng Nam-Đà Nẵng* newspaper account, by May 1981, 15 collectives had experimented with product contracts. The performance of these collectives reportedly improved significantly. Collective members' responsibility for tending paddy fields had been enhanced in an 'unprecedented way', and 'everyone was daily and nightly concerned about how to exceed the quota'.[6] The report said peasants usually worked on their contracted rice fields—even on 30 (lunar) December, an important day of the *Tết* holiday (Vietnam's New Year festival) and one day before the New Year. Collectives had cultivated their fields fully and on time, prepared the land properly, transplanted

---

5   Bàn về công tác khoán sản phẩm cuối cùng đến người lao động trong sản xuất nông nghiệp [Discussing the assignment of final products to labourers in agricultural production], *Quảng Nam-Đà Nẵng*, 18 March 1981, p. 1.
6   Qua các hợp tác xã làm thử việc khoán sản phẩm cuối cùng đến người lao động [An evaluation of the performance of collectives adopting the product contract], *Quảng Nam-Đà Nẵng*, 23 May 1981, p. 1.

paddy according to the right techniques, weeded assiduously and so on. The paper reported that members had improved agricultural intensification on contracted fields and had used manure of a higher quality than previously. Many bought extra chemical fertilisers to supplement their contracted paddy fields, and staple food production, labour productivity and yield had increased substantially.[7]

Excited with the good performance of collectives in adopting product contracts, in early July 1981, QN-ĐN chairman Phạm Đức Nam called for an intensification of the contract system. He said the product contract was correct policy that met collective members' aspirations and needs and helped enhance further collectives' economic performance. As well as contributing to a 40 per cent increase in paddy productivity, he argued, the product contract system had helped strengthen collective organisations, especially those that had been close to collapsing. In addition, product contracts helped improve collective management and the fight against negativism—rather than reviving peasants' consciousness of individual farming, as some critics had worried.[8]

On the back of these good results, in August 1981, QN-ĐN leaders called for an expansion of the use of contracts to secondary crops. They urged collectives to take back secondary-crop land, which had been temporarily lent to peasants.[9]

By the end of July 1981, 165 of the 241 collectives in QN-ĐN had adopted the product contract system; in Đại Lộc and Hòa Vang districts, all collectives implemented the policy. By the end of the winter–spring of 1981–82, all collectives in QN-ĐN had completed the adoption of product contracts.[10]

---

7   Ibid., p. 1.

8   Phạm Đức Nam: Tích cực thực hiện khoán sản phẩm cuối cùng đến người lao động trong nông nghiệp [Phạm Đức Nam: Be positive in implementing the product contract], *Quảng Nam-Đà Nẵng*, 1 July 1981, p. 1.

9   Khoán sản phẩm trên đất màu để làm vụ Đông–Xuân tốt nhất [Making contracts on secondary-crop land for the best winter–spring crop], *Quảng Nam-Đà Nẵng*, 15 August 1981, p. 1.

10   Tổng kết 3 năm thực hiện khoán sản phẩm đến người lao động trong nông nghiệp (1981–1984) [Summary of three years of implementing the product contract (1981–1984)], *Quảng Nam-Đà Nẵng*, 6 July 1985, p. 1.

In Thăng Bình district, the implementation of product contracts was also rapid. By October 1981, all collectives in the district had adopted the system.[11] Villagers in Hiền Lộc recalled that Bình Lãnh collective adopted product contracts in 1981. Under these contracts, peasants were in charge of three phrases (*ba khâu*) of production—ploughing and harrowing, planting and tending, and harvesting—which differed somewhat from the national policy. The collective teams supplied seedlings, applied fertilisers, irrigated the fields, controlled and prevented diseases and monitored distribution after the harvest.[12]

The implementation of product contracts in many other provinces of the Central Coast and Central Highlands was also swift. In the winter–spring of 1980–81, 341 of 1,101 collectives in the Central Coast and 105 of 285 collectives in the Central Highlands began to experiment with the system.[13] Moreover, by July 1981, 53.8 per cent of collectives in these regions had adopted product contracts. At a conference on collectives in the Central Coast and Central Highlands in July 1981, party researchers claimed the product contract policy met the aspirations of people and local cadres and had significantly contributed to increased productivity in the collectives. Local 'negative practices' had also been significantly reduced. At the conference, Nguyễn Ngọc Trìu, a central government agriculture minister, also asserted that product contracts played an important role in strengthening collectives and facilitating agricultural production. He called for the completion of the adoption of product contracts in these two regions by the winter–spring of 1981–82.[14] Within one year, almost all collectives in the Central Coast had implemented product contracts.

Product contracts immediately enhanced the performance of collective farming and boosted agricultural output in QN-ĐN.[15] To illustrate these improvements, the *Quảng Nam-Đà Nẵng* newspaper

---

11   Các hợp tác xã ở Tam Kỳ, Thăng Bình, Tiên Phước căn bản hoàn thành khoán sản phẩm vụ Đông–Xuân [Collectives in Tam Kỳ, Thăng Bình and Tiên Phước have completed the adoption of the product contract in the winter–spring crop], *Quảng Nam-Đà Nẵng*, 28 October 1981, p. 1.

12   Author's interview, 19 October 2005, Hiền Lộc.

13   Ban Quản Lý Hợp Tác Xã Nông Nghiệp Trung Ương [hereinafter Ban Quản Lý HTX NN TU] (1982), Khoán sản phẩm trong hợp tác xã và tập đoàn sản xuất nông nghiệp [The Product Contract in Collectives and Production Units], Hà Nội: NXB Sự Thật, p. 65.

14   Hội nghị khoán sản phẩm trong hợp tác xã nông nghiệp ven biển Trung Trung Bộ và các tỉnh Tây Nguyên [A conference on the product contract in the Central Coast and Central Highlands collectives], *Quảng Nam-Đà Nẵng*, 8 July 1981, p. 1.

15   Phạm Đức Nam: Phát huy thắng lợi bước đầu mở rộng khoán sản phẩm cuối cùng ở tất cả hợp tác xã nông nghiệp cả tỉnh [Phạm Đức Nam: Extending the product contract to the rest of the collectives], *Quảng Nam-Đà Nẵng*, 23 September 1981, p. 11.

in 1981 printed several articles praising the positive effects of product contracts. Among these was a letter from a peasant that criticised old farming arrangements and praised the new product contract system in his village. He wrote that, previously, under the work-points system, villagers merely pretended to work. When ploughing, people did one line and skipped another. When carrying manure to the fields, they dropped a lot along the road. They did not take care of collective production but concentrated on accumulating as many work-points as possible. But now, under the product contract system, everyone took care of their contracted fields. They ploughed their land properly. They transplanted and spread fertilisers and manure according to the right techniques, and many had increased their use of fertilisers and manure—all of which showed that villagers wanted to produce beyond their quota.[16]

Quảng Nam-Đà Nẵng also reported several typical cases of collectives whose performance was significantly improved thanks to the adoption of product contracts, one of which was Tam Ngọc collective in Tam Kỳ district. Its performance during the work-points system had been very poor; in the spring–summer of 1980, the average productivity of paddy was 1.1 tonnes per hectare, meaning the collective was unable to fulfil its food obligation to the state, owing 8 tonnes of paddy. In the spring–summer of 1981, collective leaders adopted product contracts and set a quota of 1.2 tonnes per hectare. Thanks to the contracts, paddy productivity increased to 2 tonnes per hectare, exceeding the quota by 0.8 tonnes per hectare. As a result, the collective members' income increased, and the collective was able to repay its previous debt and also fulfil its state obligation of 42 tonnes of paddy. The main reason for the increase in paddy productivity, the article argued, was the product contract system, which encouraged members to care for their contracted fields more than ever before.[17]

In general, *Quảng Nam-Đà Nẵng* accounts showed that, under the product contract system, the performance of collective farming in the province had significantly improved. In 1981, for the first time since reunification, the province had produced nearly 500,000 tonnes

16   Thư xã viên: Cách khóan mới ở quê tôi [Member's letter: New method of contracts in my village], *Quảng Nam-Đà Nẵng*, 4 July 1981, p. 3.
17   Một vài cách vận dụng khoán sản phẩm cuối cùng về cây lúa ở hợp tác xã Tam Ngọc [Application of the product contract for rice fields in Tam Ngọc collectives], *Quảng Nam-Đà Nẵng*, 12 September 1981, p. 2.

of staple food, which was close to the province's own consumption.[18] Staple food production in 1982 reached 525,000 tonnes of paddy equivalent, while staple food per capita increased from 303 kilograms in 1979 to 342 kilograms in 1982.[19] Kerkvliet's study of the northern collectives showed similar improvements; thanks to product contracts, staple food production in the north in 1981 and 1982 averaged a 24 per cent increase over 1980.[20]

Table 6.1 Rice production in QN-ĐN, 1979–82

Year	Area (hectares)	Yield (tonnes)	Annual growth of yield (%)
1979	124,739	319,917	
1980	123,329	310,742	–2.87
1981	122,734	332,211	6.91
1982	123,575	347,572	4.62

Source: Diễn biến sản lượng lúa cả tỉnh qua các năm [Paddy production over the past years], *Quảng Nam-Đà Nẵng*, 14 September 1983, p. 1.

A few seasons after adopting product contracts, however, collective farming in QN-ĐN started to falter. Although the province's leaders made great efforts to strengthen them, collectives became weaker. Staple food production in QN-ĐN stagnated, especially during 1985–88, and the living conditions of collective members deteriorated. Villagers in Thanh Yên and Hiền Lộc recalled that the product contracts improved the performance of collective farming in the first few seasons, but their living conditions then deteriorated because they were unable to produce beyond the quota—a serious problem that is discussed in the next chapter.[21]

18   Mặt trận sản xuất nông nghiệp: Thành tựu của năm 1981 và nhiệm vụ vụ Đông–Xuân 1981-1982 [Agricultural production: The achievements of 1981 and the ongoing tasks for winter-spring 1981-1982], *Quảng Nam-Đà Nẵng*, 28 October 1981, p. 1; Tổng kết sản xuất nông nghiệp năm 1981 và phát động thi đua giành vụ Đông–Xuân 1981–1982 thắng lợi toàn diện, vượt bậc [Summing up agricultural production in 1981 and calling for high achievements in the winter-spring of 1981–1982], *Quảng Nam-Đà Nẵng*, 4 November 1981, p. 1.
19   Diễn biến sản lượng lúa cả tỉnh qua các năm [Paddy production over the past years], *Quảng Nam-Đà Nẵng*, 14 September 1983, p. 1.
20   Kerkvliet, *The Power of Everyday Politics*, p. 194.
21   Author's interviews, October 2004; October–December 2005, Hiền Lộc and Thanh Yên.

An Giang, in the Mekong Delta

Adopting the product contract system

The Southern Region began to experiment with product contracts in the summer–autumn of 1981—a season later than their counterparts in the Central Coast.[22] In mid-1981, An Giang's leaders called for trials of product contracts in the two production units in Bình Phú commune in Châu Thành district.[23] However, by the winter–spring of 1981–82, 180 of 394 production units and five of six collectives in the province had adopted the contracts.[24]

Like their counterparts in QN-ĐN, production units in An Giang found the adoption of product contracts improved their collective farming performance. Members of production units and collectives were 'enthusiastic' about the new contract system because of their increased income.[25] Some members who had previously doubted collective farming now had confidence in it and 'actively worked the contracted fields'. Many members 'spontaneously dug channels to ensure sufficient water for their paddy fields and overcame the fertiliser shortage by using manure or extra compost bought from the free market'.[26] Some collective members who had been fed up with the work-points system and had dropped out now returned to receive contracted fields. As one man in production unit no. 3 in Tây Khánh B hamlet (Long Xuyên) commented:

> I had previously neglected collective farming and left because I saw people mistrusting each other on every task [*nạnh hẹ nhau*]. I, a primary labourer, tried to work hard while other households sent their young children to work for form's sake. Now, under product contracts, I will not neglect farming any more.[27]

---

22 Ban Quản Lý HTX NN TU, *The Product Contract in Collectives and Production Units*, p. 65.

23 Collectivisation in An Giang has progressed, *An Giang*, 18 November 1981, p. 1; Xã luận: Ra sức phấn đấu đưa phong trào cải tạo xã hội chủ nghĩa đối với nông nghiệp ở tỉnh ta tiến lên một bước mới [The editorial: Do the best to take collectivisation in An Giang one step forwards], *An Giang*, 18 November 1981, p. 11.

24 Ban Nông Nghiệp Tỉnh Ủy An Giang: Thắng lợi của việc khoán sản phẩm trong nông nghiệp ở tỉnh nhà [An Giang Provincial Committee of Agriculture: The victory of the product contract in the province], *An Giang*, 4 July 1982, p. 3; Kết quả tốt đẹp của khoán sản phẩm trong nông nghiệp [The product contract brings about good results], *An Giang*, 23 May 1982, p. 1.

25 Phấn khởi với cách khoán mới [Enthusiasm with the product contract], *An Giang*, 14 March 1982, p. 2.

26 Vụ lúa khoán đầu tiên ở thị xã Long Xuyên [The first contracted rice crop in Long Xuyên town], *An Giang*, 14 March 1982, p. 1.

27 Cited in The results of contracted rice crops, *An Giang*, 19 April 1982, p. 3.

According to *An Giang* newspaper accounts, product contracts achieved the following results: first, food production increased, mainly because the new contracts made peasants 'enthusiastic and eager to work' (*tự giác lao động*). Paddy productivity in collective farming increased, from 2–3 tonnes per hectare in the winter–spring of 1980–81 to 4–4.5 tonnes per hectare in the winter–spring of 1981–82. For example, in Chợ Mới district, paddy productivity increased from 2.5–3 tonnes per hectare to 6–8 tonnes in the winter–spring of 1981–82.

Second, the living conditions and income of collective members increased accordingly. The amount of paddy that each hectare of rice produced beyond the quota ranged from 300 kilograms to 1.5 tonnes. In Chợ Mới district, the amount exceeding the quota per hectare increased from 1 tonne to 2 tonnes, and the value of work-points increased from 1.2–5 kilograms of paddy to 10 kilograms of paddy in the winter–spring of 1981–82.

Finally, product contracts helped improve and strengthen collective farming in production units and collectives. Land, inputs and labour were better used and incidents of land being abandoned, laziness, foot-dragging, dropping out and embezzlement had been significantly reduced. Many production units and collectives had fulfilled two-way exchange contracts (*hợp đồng hai chiều*) with the state, paying irrigation fees and old debts, while food procurement (mobilisation) and sales (*huy động lương thực*) to the state increased to the amount of 1.2–2 tonnes per hectare.[28]

Satisfied with these achievements, provincial leaders urged the expansion of the product contract system. By the winter–spring of 1982–83, 896 production units (86 per cent of the total) and six collectives (100 per cent) in An Giang had adopted product contracts.[29] In Chợ Mới district, the adoption of product contracts was also rapid.

---

28   Khoán sản phẩm cuối cùng đến người lao động, một hình thức thích hợp mang lại nhiều kết quả to lớn [The product contract is suitable and brings about good results], *An Giang*, 30 May 1982, p. 1; The product contract brings about good results, *An Giang*, 23 May 1982, p. 1; An Giang Provincial Committee of Agriculture, *An Giang*, 4 July 1982, p. 3.
29   Các tập đoàn sản xuất, hợp tác xã tiến vào vụ Đông–Xuân 1982–1983 với nhiều khí thế mới [Production units and collectives entered into the winter–spring of 1982–1983 with new enthusiasm], *An Giang*, 2 January 1983, p. 1.

By the winter–spring of 1981–82, 39 of 40 production units were using the system. The four production units in Long Điền B took up product contracts in the winter–spring of 1981–82.[30]

Villagers there recalled that, in order to take up these contracts, collectivised land in the four production units had to be randomly (*bóc thăm*) divided among each household's primary worker. Each primary worker received 1 *công* (1,000 sq m) of contracted land and a subsidiary worker received 0.5 *công* (500 sq m).[31] In May 1982, *An Giang* newspaper reported that, in the winter–spring of 1981–82, the production units in Long Điền B in Chợ Mới district had all produced a bumper harvest; the average productivity of paddy was 6 tonnes per hectare. Explaining the increased performance, the article quoted one peasant, who said:

> Frankly speaking, under product contracts, every member took care of the paddy fields, so every member had a bumper harvest. I had never seen abundant harvests like these since the beginning of production units![32]

A former cadre of production unit no. 1 said product contracts had saved some units from collapse and facilitated collectivisation in the Southern Region. The main reason was peasants' resistance to collective farming under product contracts was weaker than it had been under the work-points system.[33] However, like their counterparts in QN-ĐN, in An Giang, a few seasons later, collective farming began to flounder and became a site of struggle between the state, cadres and peasants over land, labour and other resources.

The second wave of collectivisation under product contracts

At the fifth party congress in March 1982, VCP leaders acknowledged the failure of socialist transformation in the previous five years and outlined a new socioeconomic five-year plan for 1981–85.[34] After this, the Southern Region intensified both collectivisation and land redistribution.[35]

---

30 Huyện Chợ Mới áp dụng khoán sản phẩm có kết quả [The product contract in Chợ Mới brings about good results], *An Giang*, 9 May 1982, p. 3.

31 Author's interviews, June–August 2005, Long Điền B.

32 Kết quả khoán ở Long Điền B [The results of the product contract in Long Điền B], *An Giang*, 2 May 1982, p. 3.

33 Author's interview, 17 August 2005, Long Điền B.

34 Vo Nhan Tri, *Vietnam's Economic Policy Since 1975*, p. 126.

35 Lâm Quang Huyên, *The Land Revolution in South Vietnam*, p. 193.

Like many provinces in the Southern Region, in An Giang, from 1982 to 1985, collectivisation was more extensive and more rapid than in previous periods. For example, from 1979 to 1981, Phú Tân district established only 30 production units. However, using product contracts, the district established 40 units during the first six months of 1982. The main reason for this acceleration was that peasants did not resist as strongly as before. In some areas, some peasants even mobilised one another to form production units.[36] A former cadre of An Giang's Committee for Agricultural Transformation asserted that product contracts made it easier to mobilise peasants into collective organisations because they were allowed to farm on their own land.[37]

An Giang newspaper accounts show that, from the issuance of product contracts, collectivisation in the province moved more quickly than before. At the end of 1981, An Giang had only 384 production units and six collectives, accounting for less than 10 per cent of peasant households and agricultural land. A year later, however, An Giang had 1,044 production units and six collectives, accounting for 34 per cent of peasant households and 19.2 per cent of agricultural land. The province had established 660 production units in 1982—double the number of units set up between 1978 and 1981.[38]

Collectivisation in An Giang and many other provinces in the Southern Region began to slow during the period 1982–83. During the first 10 months of 1983, An Giang established only 164 production units.[39] The reasons for this were unclear. It seems a shortage of cadres was one important factor. In addition, from 1982 to 1983, authorities put greater effort into strengthening the newly established collective organisations and training cadres than into expanding the number of new production units.[40] Explaining the slowing of collectivisation in

---

36   Phú Tân tiến nhanh trong phong trào hợp tác xã hóa nong nghiệp [Collectivisation in Phú Tân advances fast], *An Giang*, 8 August 1982, p. 2.

37   Author's interview, 6 June 2005, Long Xuyên.

38   Ban Tuyên Huấn tỉnh Ủy An Giang: Thành tích cải tạo nông nghiệp của tỉnh An Giang [An Giang Provincial Committee of Propaganda: The achievements of agricultural transformation in An Giang], *An Giang*, 2 January 1983, p. 1.

39   Đẩy mạnh cải tạo quan hệ sản xuất nông nghiệp [Speeding up agricultural transformation], *An Giang*, 25 September 1983, p. 2.

40   *An Giang* reports in 1983 often called for the strengthening of production units and collectives. See, for example: Toàn tỉnh đẩy mạnh cùng cố và phát triển tập đoàn [The province intensifies the solidification and extension of production units], *An Giang*, 12 June 1983, p. 1; Các địa phương tập trung công tác cùng cố, nâng chất và phát triển tập đoàn sản xuất [Local authorities must focus on solidifying, improving and extending production units], *An Giang*, 7 August 1983, p. 2.

the province in 1983, a former Chợ Mới district official argued that, after seeing collective farming's unpleasant outcomes, the provincial party leaders wanted to halt its progress. Moreover, in the early 1980s, many northern cadres who came to provide support for agricultural transformation returned home on expiry of their official duty or after conflict with local cadres.[41]

Unhappy with the slow progress of collectivisation and land redistribution in the Southern Region, the central party secretariat issued Directive No. 19/CT-TW on 3 May 1983, urging the Southern Region to accelerate the socialist transformation of agriculture and setting 1983 as the target for completion of land redistribution and 1985 for collectivisation.[42] However, neither land redistribution nor collectivisation accelerated.

In September 1983, *An Giang* newspaper reported that collectivisation in the province had been extended 'slowly and not widely in all districts. There were still 28 communes where no production unit had been established.' Provincial leaders therefore again called for an intensification of agricultural transformation and urged the completion of collectivisation by 1985 as a national target.[43] In October 1983, the provincial party committee released an urgent action plan (*chương trình hành động*), pushing each district to determine the main causes of slow collectivisation and to set about reversing them. Provincial leaders also insisted that some key districts, such as Long Xuyên, Chợ Mới and Châu Đốc, achieve complete collectivisation by 1984.[44]

According to one study, from 1984 to 1985, collectivisation in An Giang was 'extended hurriedly'.[45] The share of collectivised land in total agricultural land jumped from 20.6 per cent in October 1983 to 30 per cent in June 1984 and to 47.6 per cent in February 1985.

---

41 Author's interview, 18 June 2005, Chợ Mới.
42 ĐCSVN (2005), Chỉ thị của Ban Bí Thư số 19/CT-TW (ngày 3 tháng 5 năm 1983) [Directive of the Secretariat No. 19/CT-TW (3 May 1983)], in ĐCSVN, *Văn Kiện Đảng Toàn Tập: Tập 44, 1983 [Party Document: Volume 44, 1983]*, Hà Nội: NXB Chính Trị Quốc Gia, p. 190.
43 Speeding up agricultural transformation, *An Giang*, 25 September 1983, p. 2.
44 Hội nghị tỉnh ủy để ra chương trình hành động từ nay đến năm 1984 [Provincial party committee meeting to make a plan of action from now to 1984], *An Giang*, 23 October 1983, p. 1.
45 Tô Thành Tâm (1990), Vấn đề ruộng đất và hợp tác hóa nông nghiệp ở An Giang [Land and collectivisation issues in An Giang], *Thông Tin Lý Luận*, 8 August, p. 8.

Table 6.2 Extending collectivisation in An Giang, 1982–85

Period	No. of production units	No. of interproduction units	Number of collectives	Percentage of agricultural land collectivised	Percentage of peasant households collectivised
December 1981	384	0	6	7.00	n.a.
December 1982	1,044	2	6	19.16	34.00
October 1983	1,216	57	6	20.61	42.08
June 1984	1,633	106	7	29.70	56.80
February 1985	1,957	116	7	47.60	n.a.
May 1985	2,326	121	7	80.00	77.13
November 1985	2,607	132	7	93.00	86.00

n.a. = not available

Sources: Ban Tuyên Huấn tỉnh Ủy An Giang: Thành tích cải tạo nông nghiệp của tỉnh An Giang [An Giang Provincial Committee of Propaganda: The achievements of agricultural transformation in An Giang], *An Giang*, 2 January 1983, p. 1; Toàn tỉnh hiện có 1216 tập đoàn sản xuất, 57 liên tập đoàn, 70 tập đoàn máy nông nghiệp [An Giang now has 1,216 production units, 57 interproduction units and 70 machinery units], *An Giang*, 23 October 1983, p. 2; Khắp nơi trong tỉnh [News around the province], *An Giang*, 12 July 1984, p. 4; Toàn tỉnh thành lập được 1957 tập đoàn sản xuất, tập thể hóa 106,798 ha [An Giang has 1,957 production units, collectivising 106,798 hectares], *An Giang*, 28 February 1985, p. 1; Đưa phong trào hợp tác hóa của tỉnh nhà lên vững chắc [Advancing collectivisation firmly], *An Giang*, 7 June 1985, p. 1; An Giang hoàn thành cơ bản công tác cải tạo nông nghiệp [An Giang has completed agricultural transformation], *An Giang*, 22 November 1985, p. 1.

By February 1985, the province was halfway towards completing collectivisation. To achieve full collectivisation in time to coincide with the tenth anniversary of Vietnam's reunification, provincial leaders pushed even harder towards completion. As a result, by the end of May 1985, collectivisation accounted for 80 per cent of agricultural land—a minimum index for basic collectivisation.[46] Unsatisfied with this achievement, however, the provincial Communist Party committee issued Directive No. 17-CT on 23 July 1985, urging an even more rapid take-up of collectivisation. By November 1985, An Giang

---

46   Advancing collectivisation firmly, *An Giang*, 7 June 1985, p. 1.

had established 2,607 production units, 132 interproduction units and seven collectives, accounting for 93 per cent of agricultural land and 86 per cent of peasant households.[47]

Collectivisation in many other provinces in the Mekong Delta was also extensive during the period 1984–85. For example, by early 1984, Tiền Giang province had established 2,515 production units and 27 collectives, which collectivised 85,953 hectares of land (77.7 per cent of the total agricultural land) and 143,158 peasant households (78.2 per cent of the total).[48] By May 1985, Cửu Long province had established 17 collectives and 4,721 production units, which accounted for 76 per cent of agricultural land and peasant households. At the same time, however, there were still many communes in which no production units, and even no production solidarity teams, had been established. Even so, by 20 October 1985, Cửu Long province announced the completion of collectivisation. The province had established 18 collectives and 5,337 production units, accounting for 97 per cent of peasant households and 94 per cent of its agricultural land.[49] Similarly, by June 1985, Hậu Giang province had established 6,983 production units and 36 collectives, accounting for 86 per cent of agricultural land and 85 per cent of peasant households. By 30 September 1985, Hậu Giang announced the completion of collectivisation in most of the province, with the establishment of 7,420 production units, 219 interproduction units and 36 collectives, accounting for 93 per cent of land and 94 per cent of peasant households.[50]

By early 1984, the whole Southern Region had established 20,341 production units and 296 collectives, which accounted for 38 per cent of agricultural land and 45 per cent of peasant households.[51] By late 1985, the Southern Region, including the Mekong Delta, had largely completed collectivisation. However, a former cadre of the Committee for Southern Agricultural Transformation recalled the hasty way in

---

47 An Giang has completed agricultural transformation, *An Giang*, 22 November 1985, p. 1; Tô Thành Tâm, Land and collectivisation issues in An Giang, p. 18.
48 Lâm Quang Huyên, *The Land Revolution in South Vietnam*, p. 197.
49 Nguyễn Thành Nam (2000), Việc giải quyết vấn đề ruộng đất trong quá trình đi lên sản xuất lớn ở Đồng bằng Sông Cửu Long 1975–1993 [Resolving land issues in the process of large-scale production in the Mekong Delta, 1975–1993], PhD thesis, Đại Học Khoa Học Xã Hội & Nhân Văn, Hồ Chí Minh, p. 78.
50 Ibid.
51 Lâm Quang Huyên, *The Land Revolution in South Vietnam*, p. 196.

which provinces in the region collectivised by 'just signing the names' (*đánh trống ghi tên*) in order to complete collectivisation by 1985. Therefore, the quality of collective organisations was poor.[52]

A former production unit leader in Long Điền B recalled how, in 1983, authorities decided to build unit no. 10 in his hamlet. The authorities came to mark its boundaries (*đóng khung*) and then invited peasant households to join. Local cadres had to visit each household to persuade them to participate. The authorities declared those households who had land within the boundary had to join the unit or they would lose their land. About 70 per cent of the invited households decided to join.[53] In September 1984, *An Giang* newspaper revealed that some local cadres had commanded peasants to 'join the production units or lose their land'. The article said coercion was an effective tool for extending collectivisation, but it failed to make the collectives strong.[54] A former cadre from Long Điền B commune recalled authorities announcing the completion of collectivisation in late 1984, while many collective organisations had not even started operating. For example, 28 production units and four interproduction units in Long Điền B had been formed, but their quality varied and was generally poor (*không đúng tính chất*). Likewise, many production units in Chợ Mới were not functioning well despite the district being hailed as the first to complete collectivisation.[55]

Asked why they decided to join production units, many villagers in Long Điền B claimed it was so they could keep their land or to receive contracted land under the product contract system. Even some upper–middle peasants joined production units.[56] Asked why he decided to join a production unit with product contracts, an upper–middle peasant who had lost some of his land to redistribution in 1983 said:

> I was discontented with the policy but could not avoid joining the production unit. Because bureaucratic and subsidised policies were imposed on us, we citizens had to obey the state.[57]

---

52 Author's interview, 6 July 2005, Hồ Chí Minh. *Đánh trống ghi tên* literally means 'banging drums to get signatures'.
53 Author's interview, 15 August 2005, Long Điền B.
54 Chuyện to nhỏ: Khẩn trương nhưng vững chắc [Some issues: Hurry up and be firm in collectivisation], *An Giang*, 20 September 1984, p. 4.
55 Author's interview, 20 June 2005, Long Điền B.
56 Author's interviews, June–August 2005, Long Điền B.
57 Author's interview, 2 August 2005, Long Điền B.

A middle peasant explained:

> We could not help joining the production unit because it managed agricultural materials such as fuel, diesel, fertilisers and pesticides. If we did not join, we could not buy these things and were disadvantaged [*chịu nhiều thiệt thòi*]. We could not even buy toothpaste [from the state].[58]

In short, from 1982 to 1985, collectivisation in An Giang spread more widely and more rapidly than in earlier periods. This was partly because of the pressure from national leaders to complete collectivisation by 1985. Local cadres in the province and the wider Mekong Delta exerted themselves to achieve this goal.[59] They even used coercive measures like their Central Coast counterparts to force peasants to join the production units, regardless of the peasants' aspirations or the quality of these organisations. Because of the product contracts and allowing peasants to farm on their own land, collectivisation did not face as much resistance as it had previously.

The second wave of land redistribution

Accompanying collectivisation was a continuous process of land redistribution in the Southern Region from 1981 to 1985. On 30 April 1981, the VCP released Circular No. 14/TB-TW, which called for the resolute and rapid implementation of the policy of land redistribution in rural areas of the Southern Region. VCP leaders complained that the landless and land-poor still made up a large proportion of the rural population in the region despite land redistribution.[60] An investigation into 80 rural areas in the region in May 1981 showed that 25 per cent of peasant households were landless or land-poor, occupying only 10 per cent of available land. While rich peasants accounted for just 2.42 per cent of peasant households, they occupied 7.1 per cent of agricultural

---

58 Author's interview, 20 August 2005, Long Điền B.

59 By examining a series of VCP policies from 1981 to 1985, such as Directive No. 93 (June 1980), two circulars (no. 14, April 1981; and no. 138, November 1981), Directive No. 19 (May 1983) and so on, one can see that the party put great emphasis on completing collectivisation in the Southern Region.

60 ĐCSVN (2005), Thông báo 14/TB-TW, ngày 20 tháng 4 năm 1981: Kết luận của Ban bí thư tại Hội nghị bàn việc xúc tiến công tác cải tạo nông nghiệp ở các tỉnh Nam Bộ [Circular No. 14/TB-TW, 20 April 1981: On facilitating agricultural transformation in the Southern Region], in ĐCSVN, *Văn Kiện Đảng Toàn Tập: Tập 42, 1981* [*Party Document: Volume 42, 1981*], Hà Nội: NXB Chính Trị Quốc Gia, pp. 198–9.

land.[61] It was therefore essential, party leaders argued, to continue with the land redistribution policy, the main contents of which had already been provided in the previous directive, No. 57/CT-TW (November 1978).[62] Party leaders released Circular No. 138/TT-TW (11 November 1981) to guide the implementation of product contracts by strictly controlling land and redistributing it among households in order to make contracts with collective members.[63]

According to a report by the Committee for Southern Agricultural Transformation, many provinces in the Southern Region now intensified their land redistribution efforts, which had been all but neglected in 1980 and early 1981. This was in response to circular numbers 14 and 138 and Directive No. 100 (on product contracts). During 1982, 13 provinces in the Southern Region redistributed 54,934.5 hectares of land. Together with the 247,963 hectares reallocated during 1975–81, this brought the total area redistributed in the Southern Region during 1975–82 to 302,896 hectares.[64]

On 3 May 1983, the VCP issued Directive No. 19/CT-TW, which stressed the completion of land redistribution by 1983 and the intensification of collectivisation in the region. It reasserted the point that socialist agricultural transformation was designed to eliminate class warfare and deal with 'who triumphed over whom in the clash between capitalism and socialism'. Therefore, each province had to carry out land redistribution 'positively and completely' by appropriating land that was beyond the work capacity of each rural capitalist, landlord, rich peasant and upper–middle peasant household and sharing it with landless and land-poor households in the commune.[65] Despite the similarity of this directive to no. 57 (in November 1978), the language here was urgent and emphatic.

---

61   Nguyễn Thành Nam, Resolving land issues, p. 88; Đào Duy Huấn, Solidifying and perfecting socialist production relations, p. 36; Lâm Quang Huyên, *The Land Revolution in South Vietnam*, p. 181.
62   ĐCSVN, Circular No. 14/TB-TW, pp. 198–9.
63   ĐCSVN (2005), Thông tri của Ban bí thư số 138/TT-TW ngày 11 tháng 11 năm 1981 [Secretariat Circular No. 138/TT-TW of 11 November 1981], in ĐCSVN, *Văn Kiện Đảng Toàn Tập: Tập 42, 1981* [*Party Document: Volume 42, 1981*], Hà Nội: NXB Chính Trị Quốc Gia, p. 443.
64   BCTNNMN, *Report on Land Redistribution*, pp. 12, 18.
65   ĐCSVN, Directive of the Secretariat No. 19/CT-TW, p. 192.

According to the Committee for Southern Agricultural Transformation's report, after the release of Directive No. 19, every province in the Southern Region stepped up implementation of land redistribution:

> [They] were resolute to complete land redistribution by 1983 to meet the target of the central leaders' policy. In 1983 the Southern Region had readjusted 72,779.8 hectares of land … [which was] equal to the total amount of land redistribution from 1979–1981. So, by the end of 1983 the whole region had readjusted about 375,677.24 hectares.[66]

According to Đào Duy Huấn, from 1983 to 1985, the whole Southern Region had redistributed 186,286 hectares. By late 1985, it had largely completed the process. In total, in the period 1975–85, some 489,183 hectares of land had been reallocated.[67]

In assessing land redistribution, the Committee for Southern Agricultural Transformation's 1984 report admitted some mistakes had been made despite great achievements in eliminating rural exploitation, strengthening the revolutionary authorities and boosting collectivisation. First, land redistribution had been uneven across different parts of the Southern Region. For example, some local authorities had carried out land redistribution completely, while others had not. In many mountainous and single-crop areas, local authorities had not implemented land redistribution at all.

Second, many local authorities had implemented land redistribution hastily without distinguishing between different types of land recipients and land-givers. They often appropriated land and redistributed it equally (cào bằng) among each worker in the commune. Such distribution harmed 'the interests of a large number of middle peasants and caused disunity among peasants and conflicts in rural areas, manifest in peasants' complaints and petitions [thưa kiện]':

> [Finally], due to the low level of socialist consciousness, some cadres and party members were not able to distinguish between labourers and exploiters. Some wanted to retain individual farming and exploitative economic activities. Some took advantage of their positions to capture public and readjusted land, good and fertile land, for themselves and to exploit peasants.[68]

---

66 BCTNNMN, *Report on Land Redistribution*, p.18.
67 Đào Duy Huấn, Solidifying and perfecting socialist production relations, p. 37.
68 BCTNNMN, *Report on Land Redistribution*, pp. 22–3.

The amount of land redistributed in the Southern Region in 1975–85 was less than the 564,547 hectares the Việt Minh had reallocated there between 1945 and 1954.[69] It was much more than the 245,851 hectares Ngô Đình Diệm's government redistributed during the period 1955–63, but less than half of the approximately 1 million hectares the land reforms of Nguyễn Văn Thiệu's government brought to tenants during 1968–74.[70]

These previous land reforms boosted commercial agricultural production, whereas the 1975–85 redistribution weakened it, by reducing the productive capacity of households classified as middle, upper–middle and rich peasants. In the late 1980s, when facing severe food shortages, production stagnation and emerging conflicts over land in the Southern Region, the VCP recognised there were shortcomings in existing land policies, especially Directive No. 19 (3 May 1983), which equalised (*cào bằng*) the distribution of landholdings among rural households without taking into account the capacity and occupation of each household. 'Therefore, commercial agriculture in the Southern Region had been set back one step.'[71] In addition, the land redistributed in 1975–85 was then farmed collectively—a method that was usually less productive than individual household farming.

The following section discusses in more detail how the second phase of land redistribution took place in An Giang province.

In An Giang

On 5 June 1982, An Giang's leaders released Directive No. 44, calling for the acceleration of land redistribution, which had been largely neglected in 1981 and early 1982, and which they planned to complete by 1983.[72] As a result of this directive, according to newspaper accounts, many districts in the province intensified land redistribution. For example, by June 1982, Châu Thành and Châu Phú districts had redistributed

---

69   Lâm Quang Huyền, *The Land Revolution in South Vietnam*, p. 25.
70   The figures for the governments of Ngô Đình Diệm and Nguyễn Văn Thiệu include a small but unknown amount of redistributed land in the Central Coast (Prosterman and Riedinger, *Land Reform and Democratic Development*, p. 139).
71   Ban Tuyên Huấn Trung Ương (1988), *Đảng trả lời nông dân một số vấn đề cấp bách về ruộng đất* [*The Party's Response to Urgent Land Problems*], Hồ Chí Minh: NXB Tuyên Huấn, pp. 6–7.
72   Land redistribution in rural areas, *An Giang*, 6 September 1982, p. 4.

2,557 hectares of land. Of these, only 500 hectares were granted to land-poor and landless households; the remainder was used by the state and district farms or was 'borrowed' by local cadres to produce food.[73]

From reunification until September 1982, An Giang had reallocated 39,157 hectares of land to 51,818 land-poor and landless households. Nevertheless, a large number of land-poor and landless households remained. In areas where land growing one rice crop per year had not been converted to double cropping, the percentage of land-poor and landless households was particularly high—about 21.6 per cent.[74]

In response to Directive No. 19 from the central government, authorities in An Giang again agitated for speedy land redistribution. In October 1983, realising that 18 per cent of households were still land-poor or landless, the provincial party committee repeated its call to meet the deadline set for the end of 1983.[75]

Despite many districts implementing land redistribution, the results fell short of expectations and, by late 1983, the province had not yet completed the task. On 12 December 1983, An Giang's provincial standing party committee released Resolution No. 05, calling for the equitable redistribution of land among members in each commune (điều chỉnh theo định xuất đất toàn xã). That policy seemed to be at odds with the VCP's policy, which referred only to reallocating appropriated land equally among poor peasants.[76]

The reallocation of land equally among commune members was carried out not only in An Giang, but also in several Southern Region provinces, such as Tiền Giang, Cửu Long and Kiên Giang. Here, land redistribution was implemented at the same time as the establishment of production units; land within the boundaries of production units was distributed equally among members regardless of their work capacity. Authorities

---

73   Trong tháng 6 phát triển 39 tập đoàn sản xuấ: Tỉnh hiện có 474 tập đoàn [In June, 39 production units were established: The province now has 474 units], *An Giang*, 11 July 1982, p. 1.
74   Land redistribution in rural areas, *An Giang*, 6 September 1982, p. 4.
75   Văn Phòng Tỉnh Ủy An Giang: Tiếp tục điều chỉnh ruộng đất củng cố và phát triển tập đoàn sản xuất [An Giang Provincial Committee Office: Continuing land redistribution, and the solidification and extension of production units], *An Giang*, 9 October 1983, p. 1.
76   Ban Nông Nghiệp Tỉnh Ủy: Tình hình điều chỉnh và qui hoạch ruộng đất ở xã Vĩnh Phú [Provincial Agriculture Board: Land redistribution in Vĩnh Phú commune], *An Giang*, 9 August 1984, p. 3.

allowed landed households to retain part of their land according to the number of people in their households, but appropriated the excess for reallocation to land-poor and landless families.[77]

In 1984 authorities in An Giang executed land redistribution extensively, to the tune of about 10,000 hectares.[78] From 1975 to July 1984, the province had appropriated 57,594.8 hectares and redistributed 56,778.9 hectares to 71,756 landless and land-short households.[79] By 30 April 1985, the province had largely completed land redistribution, having redistributed 60,225 hectares to 75,558 landless and land-poor households since 1975. Therefore, redistribution in An Giang had affected about 27 per cent of the total agricultural land (224,357 hectares) and benefited 32 per cent of the province's 233,612 peasant households.[80] According to provincial documents, in 1975, 40 per cent of peasant households in An Giang were landless or land-poor. Land reform from 1975 to 1985 therefore brought land to 81.05 per cent of the targeted beneficiaries.

However, it was later discovered that a large amount of redistributed land did not go to poor peasants but fell into the hands of local cadres. This angered peasants, and several sent petitions to authorities at all levels. For example, in July 1987, a local newspaper reported that authorities in Thạnh Mỹ Tây commune had implemented the land policy incorrectly and redistributed land irrationally (*bất hợp lý*):

> Most of the people in Thạnh Mỹ Tây were discontented with the results of land redistribution because they considered it [was] based on individual sentiment and injustice and because it favoured commune and district cadres' families.[81]

---

77   BCTNNMN, *Report on Land Redistribution*, p. 17.

78   Toàn tỉnh thành lập được 1957 tập đoàn sản xuất, tập thể hóa 106,798 ha [An Giang has 1,957 production units, collectivising 106,798 hectares], *An Giang*, 28 February 1985, p. 1.

79   Khắp nơi trong tỉnh [News around the province], *An Giang*, 12 July 1984, p. 4; Xã luận: Đẩy mạnh cải tạo quan hệ sản xuất nông nghiệp [The editorial: Intensifying agricultural transformation], *An Giang*, 9 August 1984, p. 1.

80   Toàn tỉnh đã xây dựng được 2570 tập đoàn sản xuất, 7 hợp tác xã và 21 liên tập đoàn sản xuất [The province has established 2,570 production units, 7 collectives, and 21 interproduction units], *An Giang*, 2 August 1985; An Giang has completed agricultural transformation *An Giang*, 22 November 1985, p. 1.

81   Còn thắc mắc về việc điều chỉnh ruộng đất ở xã Thạnh Mỹ Tây [Some concerns about land redistribution in Thạnh Mỹ Tây commune], *An Giang*, 31 July 1987, p. 6.

People claimed that cadres rather than average people were the beneficiaries, and that cadres accumulated large amounts of land for themselves and then lent it to others. Meanwhile, many poor households received low-quality or inadequate amounts of land, while some did not receive any at all.[82]

In retrospect, the *An Giang* newspaper concluded, in 1987:

> In past years, the implementation of land redistribution has not been correct. Some cadres, especially local cadres, took advantage of their power [*lạm dụng chức vụ*] and gave themselves, their relatives and families good land. Some local cadres did not directly cultivate but tried to accumulate land. Many state agencies at provincial and district levels also made use of their collective names to misappropriate land. On the other hand, due to the constraints of administrative mechanisms [such as prohibiting non-resident cultivators], a large amount of land was abandoned ... This led to many peasant households not having land or enough land for their production.[83]

Villagers in Long Điền B commune recalled two types of land redistribution between 1975 and mid-1986. The first was done according to the 'sharing one's clothes and rice' policy, which took place from 1975 to 1981 (before the adoption of product contracts). The second was allocating an equal amount of land to each labourer in the commune (*chia theo định suất*), which began after the adoption of product contracts and continued until 1986.[84]

Under the product contract system, land in four collective farming production units in Long Điền B was initially divided equally among collective members. In addition to four production units, commune authorities often combined the implementation of directive numbers 100 and 19 with the establishment of new production units. For example, when establishing a production unit, land within its boundary was supposed to be distributed equally among its members. In practice, land-surplus households had the right to retain any part of their

---

82   Ibid., p. 6.
83   Ý kiến: Không nên ngộ nhận giữa việc phân bố chia cấp đất đai cho hợp lý với việc trả lại ruộng đất cho chủ cũ [The opinion piece: Don't mistake rational reallocation of land for returning land to previous landowners], *An Giang*, 29 May 1987, p. 1.
84   Author's interviews, June–August 2005, Long Điền B.

land equal to the amount allocated to everyone else. The surplus was distributed to land-poor and landless households. One middle peasant recalled:

> I had 15 *công* of land. After readjusting according to land rations, I only had a few *công* left. My household had five workers, so I only retained 5 *công*. We selected some of our land according to our land ration and surrendered the remaining land for others.[85]

An elderly couple who retained only 1 *công* of land recalled:

> We had 12 *công* but almost all of our holdings were destroyed [*phá tan hết*]. At that time [about 1983–84], the state made all decisions without listening to us. Anyone was granted land; the poor and the rich had the same amount of land. A woman aged above 55 years old and a man above 60 each received only 0.5 *công* of land. The children received the same.[86]

To please landowners, local cadres in Long Điền B often allowed land-surplus households to redistribute some of their land to their relatives and acquaintances. A former cadre of interproduction unit no. 3 (*liên tập đoàn 3*) remembered that, by allowing landowners to redistribute their land to relatives, there was less resistance to the redistribution. He only readjusted whatever land remained beyond what each household had been able to disperse.[87] A former leader of production unit no. 16 said he was able to complete land redistribution in his unit in just two days (in 1984). He organised a meeting and told landowners with substantial holdings to distribute their land to relatives but not to anyone else. They all agreed and the redistribution was rapid.[88] Similarly, a former leader of production unit no. 13 said:

> At the time of production solidarity teams [1978–84], the amount of land owned still varied among households. But at the peak time of production units [1984 and 1985], land was divided equally among members according to land per capita in the commune. My method was to let the land-surplus households redistribute their land to their family members and relatives. They had the right to retain the best

---

85 Author's interview, 2 August 2005, Long Điền B.
86 Ibid.
87 Author's interview, 20 June 2005, Long Điền B.
88 Author's interview, 11 August 2005, Long Điền B.

land. Then the production unit readjusted the remaining surplus land. We conducted redistribution in this way in order to avoid hurting [people's] feelings [*tránh khỏi mất lòng*].[89]

In general, from 1982 to 1985, land redistribution in Long Điền B achieved better results and faced less resistance than during previous periods—for several reasons. First, many years after reunification, local authorities were strengthened significantly because the number of local cadres had increased and they had been better trained. Moreover, under strong pressure from higher-level authorities to complete agricultural transformation by 1985, local cadres resorted both to using harsh measures to coerce peasants and to modifying state policies to ease their resistance. For example, peasants in Long Điền B remembered that, when implementing land redistribution, authorities often sent to the fields armed cadres who were ready to arrest anyone who dared to openly resist the policy.[90] A local newspaper in 1985 printed a peasant's letter complaining that authorities in Long Kiến commune in Chợ Mới district had taken advantage of their power to redistribute peasants' rice fields, which were under cultivation, and had handcuffed people who tried to prevent cadres from redistributing land.[91]

Second, after several campaigns of socialist transformation in rural areas—including agricultural, trade and industrial transformation— in the mid-1980s, the economic power of large landowners had been weakened significantly. In addition, the state forced the adoption of high-yielding rice, the production of which depended heavily on state inputs. Realising they were not able to farm all their land on their own, many land-surplus households gave some of their land to others.[92]

Finally, despite some poor households refusing to receive others' land, local authorities faced fewer problems redistributing the appropriated land thanks to a large number of land recipients. Land recipients were not only landless or land-poor peasant households, but also included non-farming people who had previously made their living as small

---

89   Author's interview, 20 August 2005, Long Điền B.
90   Author's interview, 27 June 2005, Long Điền B.
91   Trả lời bạn đọc: Về việc điều chỉnh ruộng đất ở xã Long Kiến [Reply to reader's letter: On land redistribution in Long Kiến Commune], *An Giang*, 27 September 1985, p. 3.
92   Author's interview, 7 August 2005, Long Điền B.

merchants, in transport, as handicraft-makers and so on. Due to the socialist transformation of trade and industry, these people returned to their hamlets to receive land. One woman land recipient recalled:

> At the time of land redistribution, many people wanted to receive land because they feared that, if they did not have land, they would be sent to the new economic zones. So, we accepted the land although we knew that a few *công* was not enough for us to make a living. We thought we could combine farming with working for wages [*làm mướn*].[93]

Another woman in the same commune also asserted that some non-farming households had accepted land because they feared being taken to the new economic zones.[94] An official of Chợ Mới district's Department of Agriculture and Rural Development who was familiar with post-1975 land redistribution commented:

> An Giang was one of the provinces in the Mekong Delta which implemented forceful socialist agricultural transformation. Under the Directive No. 100, An Giang had peasantised all the rural population [*nông dân hóa mma người*]. Bike-taxi riders [*xe thồ*], pedicab riders [*xe xích lô*] and small merchants in Chợ Mới town were put into production units to receive land. So, landholdings became fragmented. This led to the poor performance of agriculture.[95]

However, some peasants—especially those who had lost large amounts of land due to the prohibition on non-resident cultivators—refused to accept other people's land. A well-off man in Long Điển B recalled that he had 60 *công* in Thoại Sơn district but lost it due to the state prohibiting non-resident cultivators. The production unit there asked him to accept a few *công*, but he refused, considering it inappropriate to take others' land when the owners were unhappy and crying. He decided to work for wages instead.[96] Similarly, a woman who had lost 130 *công* in the late 1970s due to the non-resident prohibition also refused to accept redistributed land. She said:

---

93   Author's interview, 3 August 2005, Long Điển B.
94   Author's interview, 5 August 2005, Long Điển B.
95   Author's interview, 23 June 2005, Chợ Mới.
96   Author's interview, 7 August 2005, Long Điển B.

> I did not receive readjusted land because I was afraid of making the landowners unhappy. Like us, they had suffered a lot to accumulate land. It was not right to take others' land. Rather, we worked for wages. Later I borrowed 10 *công* of my sister's land to make a living.[97]

Despite the theory that land redistribution in Long Điền B meant equal distribution among households, this was not how it occurred in practice. One peasant commented that local authorities did not appropriate surplus land from their relatives or from powerful cadres (*người có chức có quyền*), but they redistributed every bit of land of those who were powerless.[98] Land recipients were forced to accept poor and unproductive land, while local cadres took good-quality, productive land for themselves and their relatives (a practice detailed in the next chapter).[99]

In short, the second round of land reform in Long Điền B and An Giang fulfilled the targets of weakening large landowners and 'peasantising' rural people; however, due to misuse of their position, local-level cadres redistributed land in unexpected ways, meaning their relatives were the beneficiaries rather than landless and land-poor households.

Conclusion

In the first few years after the adoption of the product contract system, the performance of collective farming improved significantly, not only in QN-ĐN in the Central Coast, but also in An Giang in the Mekong Delta. As a result, the contract system was welcomed by the members of collective organisations and adopted extensively in both provinces.

In An Giang, the use of product contracts and allowing households to retain some of their land meant authorities were able to accelerate collectivisation and land redistribution. However, with strong pressure from central party leaders to complete collectivisation by 1985, authorities in An Giang and elsewhere extended collectivisation and land redistribution too hastily, while modifying the content of the national policy to speed up the process, especially during 1984–85.

---

97 Author's interview, 10 August 2005, Long Điền B.
98 Author's interview, 2 August 2005, Long Điền B.
99 Author's interview, 12 August 2005, Long Điền B.

By late 1985, those provinces had largely 'completed' collectivisation, but many collective organisations were unstable or existed only on paper.

The completion of collectivisation, egalitarian land redistribution and other policies from 1981 to 1985 changed major features of An Giang's agriculture, from commercial farming to subsistence-oriented farming, as they did in the Central Coast. They eventually destroyed the diverse rural economy of An Giang and many other areas of the Southern Region. From that time—especially from mid-1985 to the late 1980s—agricultural production in the Southern Region stagnated and then declined, peasants' living standards dropped alarmingly and a new class of exploiters started to emerge—in contrast with the VCP's original vision. In other words, from 1985 to the late 1980s, collective farming in both An Giang and QN-ĐN was in crisis, with similar serious problems and local politics in both places, which led not only local authorities, but also the VCP leadership to rethink the direction and purpose of collective farming and, finally, to decide to return to household-based farming.

7

Local politics and the withering of collective farming, 1981–88

Introduction

When the Vietnamese Communist Party (VCP) released Directive No. 100 to officially endorse the product contract system, it hoped the new farming arrangements would create an incentive for the members of collective organisations to produce beyond their contracted quota, which would increase productivity and yields. VCP leaders also expected the product contracts would reduce peasants' resistance and local cadres' negativism and strengthen collective organisations.

The product contracts did in fact immediately improve the performance of collectives and boosted agricultural production in both Quảng Nam-Đà Nẵng (QN-ĐN) and An Giang for the first few years. After that, however, they lost their momentum and failed to sustain collectives' performance and deal with local malpractices. The product contracts solved some of the peasants' concerns, but not the inherent governance challenges of collective farming and the struggle and conflicts of interest between households, collectives and the state.

In QN-ĐN, peasants and many collective leaders did not conform to the product contract guidelines and instead adopted 'blank contracts' and other improper variations (*nhiều lệch lạc*). The number of collectives adopting blank contracts increased, and the collectives' debts to the state and households' debts to the collective increased annually. Fed up with their growing debt, some collective members decided to return all or part of their contracted land to the collectives so they could find off-farm jobs. Despite provincial leaders putting great effort into solidifying collective organisations during the period 1981–88, collective farming increasingly weakened in terms of collective ownership, management and production and tended to collapse, leading to a return to individual farming.

In An Giang, the product contracts aided the completion of collectivisation and land redistribution, but could not improve or strengthen collective farming. Despite several campaigns by the provincial authorities to solidify and upgrade the quality of production units and collectives, many units remained weak and did not operate according to collective principles. The adoption of blank contracts became widespread, and production unit cadres allowed peasants to undertake all phases of farming (*buông trắng cho dân*). In addition, collective cadres often took advantage of their position to embezzle or steal collective resources and oppress the masses. As a result, food production in some production units did not increase; the living conditions of their members were poor and food contributions to the state decreased. In early 1988, An Giang's leaders began to question the direction of their agrarian policy and shifted to a preference for household farming, which contributed to a major change in provincial policy in 1988.

This chapter will continue to examine local politics, especially the forms and magnitude of QN-ĐN and An Giang peasants' and local cadres' everyday practices during the product contract period (1981–88), and reveals how and why collective farming faced similar problems and comparable practices in both locations. It will show how that behaviour adversely affected collective farming under the product contract system. The chapter also discusses how local politics and 'illegal' arrangements contributed to derailing and shifting the Communist Party's agricultural policy from collective farming to individual farming.

Local politics during 1981–88 in QN-ĐN, Central Coast region

Peasants' everyday politics

Household economy versus collective economy

Soon after the experiments with product contracts began, the *Quảng Nam-Đà Nẵng* newspaper mentioned 'a tough struggle over bad thoughts' (*cuộc đấu tranh tư tưởng phức tạp*) in the countryside, as peasants tried to harmonise their interests with those of the state and the collectives. Local authorities still insisted that, under the product contract system, collective members' earnings would come mainly from the value of their collective workdays (*giá trị ngày công tập thể*) and partly from the amount each household produced beyond their quota. In reality, many peasants were dubious about this and expected the opposite. Therefore, many wanted collective organisations to lower the quotas on their contracted fields to increase the income coming from their individual farming efforts.[1]

Despite being encouraged by the authorities to care for collective interests, many peasants mainly looked after their own household economy (*kinh tế gia đình*), where they saw a direct link between their efforts and the rewards. For example, according to the *Quảng Nam-Đà Nẵng* newspaper, in the summer–autumn of 1981, when ripe paddy fields in some collectives in northern parts of the province were suddenly flooded, many collective members took advantage of the situation and asked managerial boards to lower their quota. Otherwise, they would refuse to harvest. As a result, many paddy fields were not harvested in time and collectives suffered huge losses.[2] Likewise, according to a former brigade leader of Bình Lãnh collective, during the product contract period, collective members and cadres continually argued

---

1   Các hợp tác xã khẩn trương thực hiện thử khoán sản phẩm cuối cùng cho người lao động [Collectives must hurry in implementing the product contract], *Quảng Nam-Đà Nẵng*, 22 April 1981, p. 1; Qua các hợp tác xã nông nghiệp làm thử việc khoán sản phẩm cuối cùng đến người lao động [Results of experimenting with the product contract], *Quảng Nam-Đà Nẵng*, 12 May 1981, p. 2; Nhìn vào đồng ruộng tập thể: Lại chuyện chung và riêng [Looking at collective fields: Collective interest versus individual interest], *Quảng Nam-Đà Nẵng*, 23 May, p. 2.
2   Nâng cao chất lượng khoán sản phẩm trong sản xuất Đông–Xuân [Improving the product contract in the winter–spring crop], *Quảng Nam-Đà Nẵng*, 25 November 1981, p. 1.

about categorising contracted land and determining quotas. Peasants wanted to have fields with modest quotas and refused to accept fields on which they were not able to produce more than the quota.[3]

Villagers in Hiền Lộc and Thanh Yên recalled that, a few seasons after the implementation of product contracts, many of them lost their enthusiasm for collective farming because they could not produce more than the quota and because the value of collective work-points was low.[4] Explaining the poor earnings villagers received from collective farming under product contracts, a former chairman of Bình Lãnh collective admitted:

> At that time, the quota was set too high. For example, the quota for 1 sào [500 sq m] of the best soil land was about 200 kilograms of paddy [equal to 4 tonnes per hectare] and the quota of work was five workdays per sào. These quotas were stipulated by the district's authorities and readjusted within five years. In general, collective members received less than 50 per cent of what they produced.[5]

The *Quảng Nam-Đà Nẵng* newspaper in March 1983 evaluated the performance of Duy Phước, one of the leading collectives in the province, and revealed that, although paddy production in the collective had increased from 2,981 tonnes in 1978 to 3,577 tonnes in 1982, the living conditions of collective members had not improved much; the value of a workday was still about 2–2.1 kilograms and 2 Vietnamese dong (VND), which was similar to that under the previous system. The reason was the costs of production were huge, accounting for 75–80 per cent of the total product. Only 43.7 per cent of production went to members; therefore, income from the collective sector was far less than peasants' expectations.[6]

Fed up with the low rewards from collective farming, peasants started to devise their own arrangements. A man in Hiền Lộc village recalled that, after the implementation of contracts, everyone had to do other

---

3   Author's interview, 24 October 2005, Hiền Lộc.
4   Author's interviews, October–December 2005, Hiền Lộc and Thanh Yên.
5   Author's interview, 24 October 2005, Bình Lãnh.
6   Hợp tác xã Duy Phước chặng đường 5 năm của phong trào hợp tác hóa nông nghiệp [Duy Phước collective over the past 5 years], *Quảng Nam-Đà Nẵng*, 5 March 1983, p. 2.

work (*làm thêm*) outside the collective sector. Some went to collect firewood and rattan to sell, while others reclaimed and worked land abandoned by the collectives.[7] One poor elderly woman remembered:

> At that time, we tried to reclaim any abandoned land on the banks of streams, small ponds, corners of contracted fields, and every little bit of land. In addition, we increased the number of crops on contracted fields. For example, on one-crop-a-year land, we grew two crops; on two-crops-a-year land, we grew three crops.[8]

Many peasants claimed they often stole some of the collective's resources, such as chemical fertilisers and pesticides, to use in their own household farming (on gardens, 5 per cent land and reclaimed land) rather than on their contracted collective fields. The *Quảng Nam-Đà Nẵng* newspaper reported in December 1983 that, in many parts of the province:

> [C]ollective members appropriated collective land for their own farming … they reclaimed new land, cleared the forest for cultivation, evaded paying taxes to the state and disobeyed the management of the collective.[9]

Meanwhile, 'collective land was cultivated poorly or abandoned'. These problems were severe in some places, especially in midland areas. The article warned that, if collective ownership was not well established soon, the increased expansion of the household economy would significantly harm the collective economy.[10]

Generally speaking, under the product contract system, the household economy expanded rapidly at the expense of collective farming. Although the household economy was officially recognised in late 1979 as an integral part of the collective economy, QN-ĐN leaders in 1984 expressed their concern about the 'transgression' of putting the household economy before the collective economy, especially in weak collectives whose managerial boards were not able to control land,

---

7 Author's interview, 24 October 2004, Hiền Lộc.
8 Author's interview, 20 October 2005, Hiền Lộc.
9 Củng cố và xác lập chế độ sở hữu tập thể trong hợp tác xã [Solidifying collective ownership], *Quảng Nam-Đà Nẵng*, 7 December 1983, p. 1.
10 Ibid., p. 1.

draught animals and labour.[11] For example, Tam Ngọc was among the weak collectives in Tam Kỳ district in which 'the struggle between two paths'—the collective (*tập thể*) and individuals (*cá thể*)—was severe. By early 1984, the Tam Ngọc collective controlled and managed only 30 per cent of collectivised land, and all draught animals had been returned to individual households. The collective was unable to control and manage labour, so each worker contributed an average of only 80–90 days of labour per year; they spent the rest of their time on their household economies. Members' income from the collective was minor, accounting for only 13 per cent of the total. Therefore, 'they did not care much about the collective economy'.[12]

According to a report from the Agricultural Department of the QN-ĐN Communist Party's Committee (*Ban Nông Nghiệp Tỉnh Ủy*), by November 1984, the household economy accounted for 70 per cent of the average farming family's income, while earnings from their collective work made up only 30 per cent. In large parts of paddy-growing lowland areas of the province where collectives were able to manage and control almost all of the land, the shares of household earnings and collective earnings in a household's total income were approximately equal. However, in the midlands, where the area of secondary-crop land was large, about 80–95 per cent of peasants' total income came from the household economy.[13] In assessing the development of the household economy during 1981–84, a provincial leader raised his concern:

> Since 1981 thanks to adopting product contracts, the potential of the household economies has been exploited well, in the form of merging [the] collective economy with the household economy. So far, the household economy has been recognised but has been loosely managed [*buông lỏng*]. So, in many locations the household economy has developed in a spontaneous, unstable and incorrect way and relied largely on free markets; in some areas where the collective economy was weak, the household economy even clashed with and encroached upon the collective economy in terms of land, labour, fertilisers and so on.[14]

---

11   The VCP had recognised and encouraged the development of the household economy since the sixth plenum of its fourth congress in 1979, and particularly after its fifth plenum in 1982 (see Vo Nhan Tri, *Vietnam's Economic Policy Since 1975*, pp. 130–1).

12   Cuộc đấu tranh giữa hai con đường đang diễn ra ở một hợp tác xã [The struggle between two paths: Cooperative and individual farming in a collective], *Quảng Nam-Đà Nẵng*, 4 January 1984, p. 4.

13   Trần Ngọc Cư-Ban Nông Nghiệp Tỉnh QN-ĐN (1984), Kinh tế gia đình ở tỉnh ta [The household economy in our province], *Quảng Nam-Đà Nẵng*, 29 November.

14   Ibid.

In response to the uncontrolled expansion of the household economy, QN-ĐN's leaders in December 1984 issued Directive No. 53/CT-TV, which, on the one hand, stressed continuing to encourage the development of the household economy, while, on the other, emphasised controlling and guiding the activities of the household economy to bring them in line with the collective economy.[15]

Despite inadequate support from local authorities, the household economy in QN-ĐN continued to rise, especially after the VCP launched the 'Đổi Mới' economic reforms of 1986, which officially recognised the existence of non-socialist economic sectors, liberalised trading and allowed freer flows of capital and labour that created more job opportunities outside the collectives. A former brigade leader in Thanh Yên village recalled:

> In the late 1980s, especially after Đổi Mới, the many young peasants abandoned or returned part of their contracted land to the collective so that they could earn a living outside the collective. Some went gold digging and some went trading. These people often got higher income than those clinging to the land. Therefore, many wanted to leave collective farming [muốn bỏ chạy ra ngoài].[16]

Similarly, a former brigade leader in Hiền Lộc village remarked that, when the country's economy was opened (mở cửa), young people left the village to earn a living elsewhere. Some worked for state enterprises, while some went to Dak Lak province in the Central highlands, where the coffee industry was booming.[17]

A former building worker (thợ hồ) in a specialised team in Bình Lãnh collective recalled that, during the later stages of the product contract system, he did not want to work for the collective simply because the value of a workday there was about 1 kilogram of paddy. Meanwhile, working for individuals, he received 3 ang (12–15 kg) of paddy per day.[18] A July article from Quảng Nam-Đà Nẵng newspaper also revealed:

---

15   Ban Thường Vụ Tỉnh Ủy QN-ĐN (1984), Nghị quyết 53/CT-TV về việc tiếp tục khuyến khích phát triển kinh tế gia đình [Provincial Resolution No. 53/CT-TV on Continually Facilitating the Household Economy], 20 December, Tam Kỳ: Quảng Nam-Đà Nẵng.
16   Author's interview, 9 December 2005, Thanh Yên.
17   Author's interview, 23 October 2004, Hiền Lộc.
18   Author's interview, 20 October 2005, Hiền Lộc.

The biggest problem [the collective faced] was that peasants in Bình Lãnh wanted to escape collective farming [*thoát ly hợp tác xã*]. By June 1987, at least 160 young people refused to accept contracted land or join specialist teams; this figure was on the rise ... The reason was that the value of [a] collective workday in Bình Lãnh collective was about 1.35 kilograms of paddy; the share of the collective economy in [a] collective household's total income was nearly 30 per cent. The excess beyond the quota was small ... [therefore,] there were two trends in peasants' behaviour. First, peasants wanted to receive less land so that they were able to intensify farming to exceed the quota. This gave them more time to care for their household economies. Second, some people, especially young people, wanted to earn a living in towns and cities because they thought collective farming could not benefit them.[19]

In short, product contracts reduced the practice of workers simply going through the motions of collective farming to accumulate work-points. They were unable, however, to motivate peasants to maximise their efforts to enhance the performance of collective farming. Rather, peasants were mainly concerned with their own household economies. Therefore, collective farming under the product contract system became a site of struggle between peasants' and collectives' interests. Peasants always tried to take advantage of any opportunity or available resources to increase their household economies, which severely harmed the collective economy.

Debt

In the first few years after the implementation of product contracts, the living conditions of collective members and food production in QN-ĐN had improved somewhat. However, at later stages of the system, collective members' living conditions stagnated or, even worse, some peasants fell into debt to the collective and the state, which they refused to pay.

According to the *Quảng Nam-Đà Nẵng* newspaper, by 1985, most collectives in Thăng Bình district owed the state because members continually postponed paying (*dây dưa*) or refused to pay what they owed to collectives. Households in the district owed about 800 tonnes of paddy. One reason was that cadres only loosely managed harvests and produce—for example, in 1985, in Ha Lam collective no. 1, each household was allowed to harvest individually. After harvesting, some

---

19 Suy nghĩ về Bình Lãnh: Sự giàu có còn ở phía trước [Think of Bình Lãnh: Prosperity is still ahead], *Quảng Nam-Đà Nẵng*, 23 July 1987, p. 3.

households used the produce for their own consumption or sold some to meet their other daily needs rather than paying their quota and other obligations to the collective. As a result, 250 of 580 households had debts totalling 130 tonnes of paddy, accounting for 12 per cent of the total collective yield.[20]

A 1987 investigation (*điều tra*) found that, despite authorities' increasing investment in agricultural inputs, staple food production and food procurement had not grown accordingly. Meanwhile, members' debts to collectives and the collectives' debts to the state had increased. For example, in the period 1984–86, collectives in QN-ĐN owed the state 25,792 tonnes of paddy equivalent. The total debt in 1986 alone was 11,903 tonnes—equal to about 4.1 per cent of the grain production in the province that year.[21]

According to accounts in *Quảng Nam-Đà Nẵng*, there were several reasons for the increased debt. First, cadres classified land and determined quotas irrationally, inaccurately and unfairly. Second, collectives did not fully service some farming phases, such as irrigation, the supply of fertilisers and the application of pesticides. Instead, they made 'straight contracts' (*khoán thẳng*) or 'package contracts' (*khoán gọn*) with collective households. Third, collectives were not able to provide resources and services to members on time, in the right quantity or of adequate quality. Therefore, many households, especially those in areas where irrigation was unreliable, suffered losses that led to them accumulating debts to the collective. Fourth, the supply of state inputs to collectives was hampered by bureaucratic red tape—for example, the level of input for a collective was calculated according to the area of land rather than actual needs. Fifth, the state set the terms of trade between agricultural inputs and agricultural produce to favour the former. Finally, authorities levied dozens of different payments from collective households, such as public bonds in paddy (*công trái thóc*), savings in paddy (*tiết kiệm thóc*), funds to help people being affected by storms and floods, paddy for training soldiers (*thóc luyện tập quân sự*), and so on.[22]

20 Tại sao tiến độ huy động lương thực ở Thăng Bình chậm? [Why is food procurement in Thăng Bình slow?], *Quảng Nam-Đà Nẵng*, 7 December 1985, p. 3.
21 Điều tra nợ lương thực: Vấn đề giải quyết lương thực hiện nay [Investigation of food debt: How to deal with food problems], *Quảng Nam-Đà Nẵng*, 29 October 1987, p. 2.
22 Củng cố và hoàn thiện công tác khoán sản phẩm cuối cùng đến người lao động trong nông nghiệp [Improving and perfecting the product contract], *Quảng Nam-Đà Nẵng*, 25 October 1986, p. 1.

Villagers in Thanh Yên and Hiển Lộc villages recalled that product contracts did not make their lives much better. Most households owed paddy to the collective. They argued that the quota was often set inaccurately—for example, in areas where production conditions were unfavourable, collective cadres still made contracts with relatively high quotas. On land unsuitable for growing rice, cadres still forced households to grow rice. Therefore, many suffered losses and fell into debt. A former chairman of Bình Lãnh collective asserted that, by the end of the product contract system (mid-1988), households owed 500 tonnes of paddy to the collective and more than 70 per cent of peasant households were in debt to the collective.[23] A man in Hiển Lộc village told of his sad experience with product contracts:

> My family received 1 *mẫu* and 5 *thước* of land [5,166 sq m]. The collective coerced us to accept a large tract. My family had only three people: my mother, my elder sister and me. At that time I was 17 years old. Among the contracted plots was 4 *sào* of land without irrigation (one crop per year). For this 4-*sào* land, I could produce only 30 *ang* [150 kg] of paddy, but had to pay the collective 480 kilograms. I ask you, how could I pay? I had to owe the collective. At that time I often went to do collective work and accumulated a large number of work-points. So, the collective used my work-points to reduce my debts but the remainder was still large. I had to plant sweet potatoes and cassavas to pay the remaining debt. We had a difficult life and always owed the collective. Some households owed tonnes of paddy. I did not understand why the collective set the quota so high. A *sào* of land without irrigation had a quota of 120 kilograms. How could we produce that amount? We knew that they had suppressed us but we did not know what to do. Some people cried a lot and begged the collective to take back the land but that hardly succeeded.[24]

Like their counterparts in Hiển Lộc village, about 90 per cent of the households in Thanh Yên village owed the Bình Định collective. By the late 1980s, the whole village (60 households) owed the collective about 22 tonnes of paddy. Some households owed more than 1 tonne of paddy each.[25] Villagers commented that, under contract number 100, paddy productivity was low due to inadequate levels of fertiliser. Collectives set unfair quotas and charged many fees and funds that were converted into paddy, so many households ended up owing tonnes of paddy

23   Author's interview, 24 October 2004, Bình Lãnh.
24   Author's interview, 21 October 2005, Hiển Lộc.
25   Author's interview, 9 December 2005, Thanh Yên.

to the collective.[26] An elderly man in the village recalled that, under the product contract system, people did not want to receive much contracted land for fear of falling below the quota and because they had to contribute more than 70 per cent of their yield to the collective. The payment for fertilisers (received from the collective) alone was half the yield.[27] Similarly, a woman whose family owed 2 tonnes of paddy to the collective—the largest single debt in the village—explained:

> Under product contracts, the collective forced us to accept large amounts of land. My family did not have cattle so we did not use manure to fertilise the fields. We did not have money to buy chemical fertilisers as others did. Moreover, my husband was fed up with collective farming and refused to work. I worked the fields alone. Therefore, we always had bad harvests and were not able to pay our debts.[28]

In response to the increased debts in the late 1980s, Bình Lãnh and Bình Định hardened their collection procedures. According to a former Bình Định collective cadre, initially, cadres relied on commune police to confiscate debtors' property (such as cattle and bicycles), but the results of such hardline property seizure were unpleasant. Later, collectives hired district court cadres and police to collect debts by paying them 20 per cent of the value of the debt reclaimed. However, results were unsatisfactory because many debtors had nothing to confiscate. Finally, authorities gave up collecting peasants' debts.[29] A former brigade leader of Bình Lãnh collective lamented:

> Many people owed the collective; they said they did not have enough food to eat so they could not pay. They said that they would pay their debts when human flesh was allowed to sell in the markets [they meant selling their own flesh to pay off debt]. They also watched each other to see if others paid or not; if not, neither did they. They copied each other [nạnh với nhau].[30]

---

26 Author's interview, 12 October 2004, Thanh Yên. Ví is a large bamboo basket that villagers often use to store paddy.
27 Author's interview, 17 December 2005, Thanh Yên.
28 Author's interview, 9 November 2005, Thanh Yên.
29 Author's interview, 9 December 2005, Thanh Yên.
30 Author's interview, 15 October 2005, Hiền Lộc.

A 70-year-old man said that, under product contracts,

> many people were in debt to the collective. If people were poor, the state failed. It was impossible for the state to kill people if they were not able to pay their debts.[31]

Abandoning or accepting less contracted land

The *Quảng Nam-Đà Nẵng* newspaper reported in late 1984 that many collectives in the province were using agricultural land wastefully and ineffectively. For example, collectives had abandoned large amounts of agricultural land or irrationally converted some of it into non-farming land. As a result, in 1984 alone, QN-ĐN had lost 5,000 to 10,000 hectares of cultivated land.[32]

A further investigation in 1985 found that another reason for the decrease in agricultural land was that some peasants had abandoned collective fields. For example, in 1984, peasants in 11 districts of the province had abandoned 13,000 hectares; in some collectives, the abandoned area amounted to hundreds of hectares.[33] Peasants abandoned collective fields largely because the costs of cultivating were too high. In November 1986, *Quảng Nam-Đà Nẵng* reported that, after adopting product contracts:

> Some collectives did not provide collective households with agricultural inputs or services sufficiently or on time, leaving them alone to take care of their crops. If collective households invested more in their contracted fields, the excess beyond the quota would not cover their expenditure ... It was worrying that collective members did not want to accept contracted land. Instead they wanted to return it to the collective in order to do outside work, which brought them higher incomes.[34]

---

31 Author's interview, 14 October 2005, Hiến Lộc.

32 Chung quanh vấn đề sử dụng đất nông nghiệp [The problem of using agricultural land], *Quảng Nam-Đà Nẵng*, 15 August 1984, p. 2; Hội nghị tổng kết sản xuất nông nghiệp, phát động chiến dịch sản xuất vụ Đông–Xuân [A conference summing up five years of agricultural production and campaigning for the winter–spring crop], *Quảng Nam-Đà Nẵng*, 30 October 1984, p. 1.

33 Hội nghị quản lý ruộng đất của tỉnh sử dụng tài nguyên đất với hiệu quả kinh tế cao nhất, chấm dứt việc cấp đất trái phép, xử lý nghiêm khắc những vụ lấn chiếm đất trái phép của nhà nước và tập thể [Provincial land management conference promotes land resources with the highest economic efficiency, terminates illegal land allocation, strictly handles illegal encroachment on state and collective land], *Quảng Nam-Đà Nẵng*, 25 April 1985, p. 1; Cần quản lý và sử dụng đất nông nghiệp một cách hợp lý [The need to use agricultural land rationally], *Quảng Nam-Đà Nẵng*, 25 April 1985, p. 2.

34 Thực sự coi nông nghiệp là mặt trận hàng đầu [Agricultural sector needs to be regarded as top national priority], *Quảng Nam-Đà Nẵng*, 27 November 1986, p. 1; Improving and perfecting the product contract, *Quảng Nam-Đà Nẵng*, 25 October 1986, p. 1.

An example of the problem was Duy Thành collective, one of the successful collectives in Duy Xuyên district during the work-points system. A few years after adopting product contracts, however, the number of households here producing beyond their quotas significantly decreased. And the more peasants invested in collective fields, the greater the losses they suffered. Therefore, 'many returned the land or kept some contracted land just as a formality [*lấy lệ*] so that they could set aside time to do other jobs to earn a living'. This led to a paradoxical situation in which the collective had a high population density but its land was abandoned.[35] Similarly, in the winter–spring of 1985–86, 30 per cent of peasant households in Điện Nam collective no. 2 in Điện Bàn district decided to return their collective land.[36] In mid-1986, for the same reason, 20 per cent of households in Bình Triều collective in Thăng Bình district returned their contracted land.[37]

Bình Lãnh and Bình Định collectives in Thăng Bình district faced problems similar to those in other districts of QN-ĐN. The number of peasants who abandoned or returned their contracted land increased annually. For example, according to *Quảng Nam-Đà Nẵng*, in 1984, hundreds of households in Bình Lãnh collective decided to accept less contracted land or return some of their poorer land to the collective. As a result, villagers were unwilling to accept contracts on 30 hectares (*không có người nhận khoán*).[38] According to a former chairman of Bình Lãnh collective, after the *Đổi Mới* reforms began, the number of households returning land to the collective increased by an average of 40–50 households per year. Some households returned land that was unfavourable for production; others returned it in order to earn a living in the Central Highlands or elsewhere. Some tried to farm on land not controlled by the collective.[39]

35   Hợp tác xã Duy Thành từng bước hoàn thiện khoán sản phẩm đối với cây lúa [Duy Thành collective gradually perfected the product contract], *Quảng Nam-Đà Nẵng*, 17 December 1987, p. 3. The collective set a quota of 8.6 tonnes per hectare per year (in the winter–spring season, 3 tonnes per hectare; the spring–summer season, 3.2 tonnes; and the third season, 2.4 tonnes).

36   Hợp tác xã Điện Nam 2 khoán mới động lực mới [New farming arrangements created new incentives in Điện Nam Collective No. 2], *Quảng Nam-Đà Nẵng*, 24 December 1987, p. 2.

37   Bình Triều 2 qua vụ Đông–Xuân 1987–1988 [Performance of Bình Triều Collective No. 2 in the winter–spring of 1987–1988], *Quảng Nam-Đà Nẵng*, 10 May 1988, p. 4.

38   Khoán sản phẩm cuối cùng đến người lao động những vướng mắc và cách giải quyết [The product contract: Problems and solutions], *Quảng Nam-Đà Nẵng*, 8 November 1984, p. 1.

39   Author's interview, 21 October 2004, Bình Lãnh.

Villagers in Hiền Lộc complained that, under product contracts, the collective established many new specialised teams (*chuyên khâu*) and other non-cultivating industries (*chuyên ngành*), which recruited many labourers. Therefore, each cultivating labourer in the collective was assigned more land than others.[40] A woman with a disabled husband and young children shared her story:

> When authorities distributed land [by drawing lots] to make contracts, I drew a lot [*bóc thăm*] of 1 *mẫu* and 7 *thước* [5,233 sq m]. That was too big for me! My husband was disabled and my children were too young; how could I manage it? I tried my best to work but contributed almost all of the produce to the collective. Thanks go to Mr Linh [Nguyễn Văn Linh],[41] who saw our problems. If collective farming had continued, I guessed that the land would be completely exhausted [because people overexploited land]. If product contracts continued for a few more years, all the people here would refuse to do collective farming.[42]

Similarly, a former brigade leader in Thanh Yên village commented that, after Nguyễn Văn Linh's ascension to power as general secretary of the Communist Party, people were freed from collective farming. He said if this had not happened, people in his village would have run away from the collective because they worked hard but received little.[43] A former chairman of Bình Định collective said:

> Under product contract no. 100, many people wanted to return their contracted land. This would raise a big question to the top leaders of why under the product contracts many people were not able to produce more than the quota; and why they wanted to abandon their collective land. In the past, landlords were thought to exploit peasants, but nobody abandoned their rented land; rather peasants often competed with each other to rent land from landlords. But now why did peasants want to abandon land?[44]

---

40 Author's interviews, October–December 2005, Hiền Lộc.
41 Ordinary people in Vietnam often attributed the *Đổi Mới* policy to former general secretary of the Communist Party Nguyễn Văn Linh. Recently, some authors have argued that Trường Chinh was the author of *Đổi Mới* (see Huy Đức, *The Winning Side*).
42 Author's interview, 15 October 2005, Hiền Lộc.
43 Author's interview, 5 October 2005, Thanh Yên.
44 Author's interview, 5 October 2005, Bình Định.

In short, everyday peasant practices such as pursuing their own household economic activities, abandoning land and accepting less collective land reflected not only their low confidence in collective farming, but also their discontent with local cadres' poor practices and mismanagement of collectives.

Local cadres' practices in QN-ĐN

In theory, product contracts diminished cadres' power and increased peasants' responsibility over the management of collective farming. Villagers were permitted to do three phases of farmwork on their own and were supposed to work collectively on the remaining five phases. However, the contract system created new opportunities for cadres to benefit at the expense of ordinary collective members. For example, managers were still given considerable power to direct collective farming, but they were not responsible for its performance. They therefore tended to shift their responsibilities on to villagers and embezzled major collective resources over which they had control, such as agricultural inputs, collective property and produce.

Quảng Nam-Đà Nẵng newspaper warned in November 1981 that it was erroneous to think the adoption of product contracts had solved all the problems of collectives. In reality, cadres in many of the collectives and brigades 'were not positive about improving management'; they did not look after the land, production tools or farmwork. Many offloaded the tasks of preparing seedlings and fertilising land to collective members without properly monitoring their performance.[45] Villagers in Thanh Yên and Hiên Lộc also recalled that cadres often failed to fulfil their duties, such as spraying to prevent insects or watering the fields on time. Therefore, seeing their paddy fields attacked by insects or short of water, villagers often tried to save their fields first rather than wait for a collective response. A former brigade leader of Bình Định collective said:

> From the outset of product contracts, the collective had nearly made 'blank contracts' with peasants; soon after implementing contract no. 100, the collective returned collective draught animals to households. In fact, the collective was only in charge of delivering fertilisers, spraying insecticides and supplying water. Meanwhile, peasants did

---

45   Improving the product contract, *Quảng Nam-Đà Nẵng*, 25 November 1981, p. 1.

everything else, but they were obliged to the collective, paying tax, agricultural input fees [for fertiliser and insecticide] and irrigation fees and contributing to collective funds.[46]

According to *Quảng Nam-Đà Nẵng*, a few years after the implementation of product contracts, many collectives increased their quotas but did not increase their investment accordingly. Some even tried to reduce production costs by lowering the price paid for manure bought from collective members and the value of a collective workday.[47] In some collectives, where cadres were allowed to farm contracted land, cadres lowered the quota to ensure they made a profit at the expense of collective earnings.[48]

When authorities ordered loans of secondary-crop land to households, collective cadres in some locations saw it as an opportunity to appropriate land for themselves and their families. When authorities wanted these fields farmed collectively again, these cadres tended to delay its return.[49] QN-ĐN's leaders considered the poor management of collectives was a result of local cadres' weaknesses. Due to inadequate training, local cadres were often incompetent and lacked discipline; some even did things at odds with state and party policies.[50]

In the period 1983–86, provincial authorities launched several campaigns to improve collectives and train local cadres, but results were below expectations. *Quảng Nam-Đà Nẵng* reported in October 1986 that collective cadres were still incorrectly implementing 'the five-farming-phase-contracts signed with collective members' by leaving households alone to do almost all the phases of farmwork; they did not conform to the requirements for managing production, inputs and outputs, which significantly affected the performance of collective farming and diminished the value of the collective workday.[51]

---

46 Author's interview, 5 November 2005, Thanh Yên.
47 Ban Nông Nghiệp Tỉnh QN-ĐN, The household economy in our province.
48 Summary of three years of implementing the product contract, *Quảng Nam-Đà Nẵng*, 6 July 1985, p. 1.
49 Solidifying collective ownership, *Quảng Nam-Đà Nẵng*, 7 December 1983, p. 1.
50 Phạm Đức Nam: Công tác trước mắt để củng cố và phát triển quan hệ sản xuất mới ở nông thôn [Phạm Đức Nam: Ongoing tasks for solidifying and improving new production relations in rural areas], *Quảng Nam-Đà Nẵng*, 18 June 1983, p. 1.
51 Improving and perfecting the product contract, *Quảng Nam-Đà Nẵng*, 25 October 1986, p. 1.

In addition, in their direction of agricultural production, collective cadres emphasised 'controlling the end-products but paid scant attention to supplying inputs, credits and technology'.[52]

After the VCP launched the *Đổi Mới* policy in 1986, the press was given more power to tackle 'social evils'. *Quảng Nam-Đà Nẵng* began to report on several problems with local cadres, including leaders of collectives. For example, an investigation in Điện Phước collective no. 2 in Điện Bàn district found 16,602 tonnes of stored paddy had disappeared during the period 1983–86; however, collective managers had not taken responsibility for this loss. Instead, they claimed 'the paddy vanished naturally [*tự nhiên biến mất*] rather than being pocketed by anyone'. They asked collective members for their understanding (*thông cảm*), and also intimidated and chastised any member who disagreed or dared to protest.[53]

Investigations also found cadres had used incorrect weight standards, thereby cheating ordinary people. Inspecting 74 scales belonging to several food-related organisations in eight of the largest paddy-producing districts in 1986, authorities discovered that only five met proper standards. The *Quảng Nam-Đà Nẵng* article said that, by weighing incorrectly, staple food officials (*ngành lương thực*) had embezzled large amounts of food in the province.[54]

Cadres in Thăng Bình district also embezzled inputs and outputs. Explaining the reasons for poor paddy productivity in the winter–spring of 1986–87 (1.6 tonnes per hectare on average, the lowest since 1976), an article in *Quảng Nam-Đà Nẵng* revealed that

> a large amount of chemical fertilisers [supplied by the state] did not go directly to collective paddy fields but passed through the hands of private merchants and then to the fields [or peasant households].

---

52 Agricultural sector needs to be regarded as top national priority, *Quảng Nam-Đà Nẵng*, 27 November 1986, p. 1.

53 Nhìn vào đồng ruộng tập thể: Hai bàn cân ở hợp tác xã Điện Phước 2 [Looking at collective fields: Two different weighing scales at Điện Phước Collective No. 2], *Quảng Nam-Đà Nẵng*, 8 August 1987, p. 2.

54 Trách nhiệm của ngành lương thực trong việc để hao hụt một số khối lượng rất lớn lương thực [State food agencies need to take responsibility for considerable loss of staple food], *Quảng Nam-Đà Nẵng*, 16 August 1986, p. 2.

In addition, many collectives merely 'fertilised on paper' (*bón phân trên giấy*). For example, investigations in five collectives in Bình Tú, Bình Sa and Bình Hải communes found that, in 1986 alone, 120 tonnes of urea fertilisers had 'flown' to the free market (*bay ra thị trường*).[55]

In 1987 authorities in QN-ĐN carried out several inspections and retrieved VND599.4 million and goods worth VND50 million that cadres had embezzled. In agricultural collectives, inspectors found 'many cases of embezzlement and theft'. For example, 'a storehouse keeper of Đại Quang collective [Đại Lộc district] embezzled 19 tonnes of paddy … [and] an interbrigades accountant of Điện Thoại No. 1 [Điện Bàn district] embezzled 35 tonnes'.[56]

In response to local cadres' negativism, the lead article in *Quảng Nam-Đà Nẵng* in September 1987 called for the widespread and full implementation of socialist democracy. It also complained:

> Local cadres and party cadres have already forgotten the lesson of 'taking people as the foundation' [*lấy dân làm gốc*] because they now lacked democratic spirit and were not close to the masses in order to hear their voices. Therefore, there were too many heart-breaking incidents such as violating the master rights of people, embezzlement and bribes which took place widely in many locations, even in some executive committees of local party organisations.[57]

Villagers in Hiền Lộc and Thanh Yên also complained that, under the product contract system, collective cadres set high quotas and raised numerous funds but used the income in ambiguous ways. Most funds went into the pockets of key collective cadres. A man in Thanh Yên village commented that 'people contributed a lot to collective funds but the collective did not do anything to benefit the people. Cadres took it all.'[58] A former brigade leader revealed that, particularly in later stages of the

---

55   Vì sao năng suất Đông-Xuân ở Thăng Bình giảm sút? [Why did rice productivity in Thăng Bình go down?], *Quảng Nam-Đà Nẵng*, 28 April 1987, p. 2; Vụ Đông–Xuân 1983–1984 Thăng Bình củng cố hợp tác xã gắn liền với tập trung chỉ đạo vùng lúa có sản lượng cao [Thăng Bình will solidify the collective and extend high-yielding rice in the winter–spring of 1983–1984], *Quảng Nam-Đà Nẵng*, 9 November 1983, p. 2; Why is food procurement in Thang Binh slow?, *Quảng Nam-Đà Nẵng*, 7 December 1985, p. 3.

56   Tổ chức thanh tra các cấp tăng cường công tác thanh tra , kiểm tra nhanh chóng phát hiện những vụ tiêu cực [Intensifying investigations of negativism], *Quảng Nam-Đà Nẵng*, 14 November 1987, p. 1.

57   Xã luận: Thực hiện rộng rãi và đầy đủ nền dân chủ xã hội chủ nghĩa [The editorial: Fully implementing the socialist democracy], *Quảng Nam-Đà Nẵng*, 1 September 1987, p. 1.

58   Author's interview, 1 October 2005, Thanh Yên.

product contract system, 'collective cadres knew that the organisation would sooner or later be dismantled so they gradually turned collective property into their own'.[59] Another former brigade leader confirmed that, after the implementation of product contracts, collective property such as tractors, waterpumps and rice-husking machines gradually disappeared, falling into the hands of collective cadres. People saw the collective property being spirited away (*hao mòn*) so they no longer wanted to contribute to the collective.[60]

The excessive number of local cadres combined with the levels of embezzlement consumed large amounts of villagers' produce. Villagers often complained that 'peasants worked, the cadres enjoyed' (*cời làm cho cối ăn*). The apparatus of collective administration consisted of so many cadres they shared among themselves much of the collective's income. For example, many collectives opened up non-farming industries, which required even more cadres to manage them—animal husbandry, brick kilns, forestry, building, carpentry and so on. All collective cadres and workers in these specialised teams had to be paid in paddy, while the non-farming income went to collective funds. Therefore, collective members' income was reduced. The collective also had to subsidise many mass organisations, which meant the collective's income was constantly being drained.[61] A former chairman of Bình Lãnh collective asserted that, under the product contract system,

> a collective was like a small state. The collective was in charge of all kinds of subsidies for local education, health care and cadre welfare. For example, when having a meeting, the district's party committee came to ask for a cow to slaughter. We had to give them one. Individual cadres from district offices also asked for help. Because the district authorities directly monitored us, when they asked for something, we had to give it to them. Commune authorities did the same. The Commune People's Committee still owes the collective about VND30 million.[62]

---

59   Author's interviews, October 2005, Thanh Yên. In November 1983, *Quảng Nam-Đà Nẵng* also reported that the privatisation of collective property through forms of 'illegal liquidation' (*thanh lý trái phép*) had begun to take place in some collectives in the province (Xã luận: Cũng cố hợp tác xã vẫn đề cấp bách đưa sản xuất nông nghiệp lên một bước [The editorial: Solidifying collectives is an urgent task to advance agriculture], *Quảng Nam-Đà Nẵng*, 14 November 1983, p. 1).
60   Author's interview, 5 November 2005, Thanh Yên.
61   Author's interview, 19 October 2005, Hiển Lộc.
62   Author's interview, 8 December 2005, Bình Lãnh.

In short, local cadres' malpractice and mismanagement contributed significantly to the poor performance of collective farming and the failure of the product contract system in QN-ĐN.

Local politics in An Giang, 1981–88

Peasants' everyday politics

Farming poorly and owing debts to production units

In An Giang province, product contracts had a brief positive effect on production units that had performed poorly under the work-points system. Product contracts also helped boost socialist agricultural transformation and bring more land and peasants into collective organisations. However, despite this and earlier land redistribution, peasants' living conditions and agricultural productivity did not improve for long. The reasons have to do with peasants' responses to collective farming.

Villagers in Long Điền B commune in Chợ Mới district recalled that land redistribution and the product contract system provided landless and land-poor households with fields to farm; however, many farmed unprofitably and ended up deeply in debt to production units. Meanwhile, some landowners who had lost land to redistribution gave up farming or grew only enough for their own consumption. A former cadre of production unit no. 9 in Long Điền B argued:

> Some guys who did not know how to farm were put into the production units to receive land. The production unit was supposed to teach them how to farm. Despite the production unit delivering fertilisers to them in advance, they did not know how to spread it properly. As you know, although the state was concerned about agricultural output, performance was low because many people did not know how to farm, while professional cultivators had lost much of their land.[63]

He also mentioned that non-farmers accounted for more than half of all land recipients in Long Điền B; most had been small traders, labourers or ran other businesses. Some lived in Cho Moi town, the capital of Chợ Mới district, and Mỹ Lương, another town in Chợ Mới. During

---

63   Author's interview, 5 August 2005, Long Điền B.

collectivisation, these people received land, but they did not know how to farm well or were unenthusiastic about farming. Therefore, after a few seasons, they often transferred, mortgaged or sold their fields to others and resumed their non-farming work. A man whose family had been river traders (*nghề ghe*) for generations recalled:

> My family had long been trading on boats so we were not good at farming. My parents previously had 3 *công* [3,000 sq m] of land but they lent it to others. At the time [about 1984], we found it difficult to continue trading on the boat because it was extremely difficult to buy fuel. Besides, it was rumoured that anyone who did not have land would be sent to the new economic zones. Therefore, I returned to farming. Because I had 3 *công* of land from my parents, I did not receive any land from others. But we farmed unproductively. My first three, four harvests were bad. I was not able to pay the fees of the B contracts [*hợp đồng B* for agricultural inputs].[64] [Consequently,] because I was not able to pay to the production unit, I was put into custody [*bị bắt nhốt*] by commune police [in 1986]. At that time, many others also owed to production units because they did not know how to farm. In addition, the production unit provided us with insufficient fertilisers, pesticides and irrigation while we contributed so much to the production units. The contribution accounted for more than half of our harvest.[65]

As well as a lack of knowhow, insufficient supplies of agricultural inputs and poorly functioning production units, many poor peasants complained they could not farm well because they were destitute and lacked capital to invest in their fields. For example, unlike the better-off peasants, poor peasants were unable to buy extra fertilisers and pesticides on the free market; they did not have enough money to hire labourers or machinery from production units to level the land, which was necessary for growing high-yielding rice.[66] Some poor villagers complained that the land redistributed to them was infertile, undulating or located in unfavourable areas. Many suffered losses and fell deeply into debt. Ultimately, some had to transfer or abandon collective land after a few seasons of farming.[67]

---

64 In the years 1979–86, at the beginning of each season, authorities (through production units) supplied peasants with agricultural inputs (fertilisers and pesticides) in return for peasants' paddy after harvest, according to the stipulated exchange rate. This arrangement was called the 'B contract'.

65 Author's interview, 12 August 2005, Long Điền B.

66 Author's interview, 2 August 2005, Long Điền B.

67 Author's interview, 5 August 2005, Long Điền B.

An investigation in Chợ Mới district in December 1985 found that 70 per cent of peasants lacked capital to invest in their farming and had begun to return land to production units.[68] This problem was widespread in An Giang. In particular, after *Đổi Mới*, when An Giang abolished some agricultural subsidies and B contracts, poor peasants faced increased difficulties in attaining agricultural inputs to invest in their farms. According to the *An Giang* newspaper in July 1987, when Châu Phú and Phú Tân districts abolished B contracts, some peasants began to abandon their contracted land because they could not buy fertiliser.[69]

Looking back, some poor villagers in Long Điền B argued that with the few *công* of land redistributed to them, their families could not live on farming alone. The income from their farming was often less than that from their previous jobs, so they had to supplement their income by doing wage work or small trading. The more time they spent working for wages, the less time they had for their own farming. Moreover, producing high-yielding rice required capital. Most poor households did not have enough capital, and production units supplied inadequate agricultural inputs. Their farming was therefore unprofitable and they ended up in debt. A former chairman of Long Điền B Commune Peasant Association shared his view:

> At first, some poor peasants were happy to receive readjusted land but later they felt dissatisfied because their farming had poor results. A poor family with five to seven people received only a few *công* of land; if all their members clung to farming, they could not survive because they could not do other business. So, they had to rely on doing wage work to supplement their livelihood. As a result, their farming was bad; their paddy productivity was about 10 *giạ* per *công* [2 tonnes per hectare; while that of better-off farmers was 4–5 tonnes per hectare]. Because they farmed inefficiently, they transferred and mortgaged [*cầm cố*] their land to others despite authorities not allowing this.[70]

---

68   Chợ Mới vào vụ mới [Chợ Mới begins to cultivate a new crop], *An Giang*, 20 December 1985, p. 2.
69   Xung quanh chuyện đầu tư cho sản xuất nông nghiệp [The problem of agricultural investment], *An Giang*, 17 July 1987, p. 2.
70   Author's interview, 9 August 2005, Long Điền B.

He also revealed that, despite the commune authorities encouraging peasants to exchange labour with each other, those in the Southern Region refused to do so, wanting instead to hire labourers rather than exchange labour.[71] This is why the rice fields of poor households often had more weeds and were unprofitable.

Similar to their counterparts in QN-ĐN, many poor peasants in An Giang could not afford the cost of agricultural inputs, taxes and other obligations to the production units and ended up in debt. Most commonly, they owed paddy for agricultural inputs (*nợ vật tư*) or B contract debts (*nợ hợp đồng B*). According to the *An Giang* newspaper, some poor peasants, after receiving inputs from production units, sold them on the free market to meet their daily needs rather than using them in their fields. They ended up with poor harvests and were unable to pay their costs. Others adopted the tactic of putting 'one foot inside and the other foot outside' the production unit so they could buy state agricultural inputs at low prices to sell on the free market at higher prices.[72] Many poor peasants were unable to pay their debts because they farmed unproductively but still had to pay production units a large amount of paddy. One poor man recalled that, during the product contract system, he always had poor harvests. If he paid fees for inputs, irrigation and his contribution to the production unit's funds, he had almost nothing left for his family. He therefore delayed his payments and was in debt to the production unit. He said: 'The authorities often came to force us to pay debts but when they saw that we were really poor, [they] finally they gave up.'[73] A former cadre of production unit no. 15 argued:

> In the time of Mr Lê Duẩn [then Communist Party secretary], it was compulsory to carry out land redistribution. But after redistributing land, because they were poor, many people farmed unproductively. They refused to pay [in paddy] fees for irrigation, fuel, fertiliser and pesticides. It was trouble. The production unit was not able to collect fees from people who had nothing. The better-off households were able to invest in their farming while the poor households just farmed. Farming like that, paddy productivity went down rapidly. The number of households who were not able to pay debts was so numerous that

---

71   Author's interview, 9 August 2005, Long Điền B.
72   Người nông dân đang cần phương thức đầu tư hợp lý phát triển sản xuất nông nghiệp [Peasants need a rational method of agricultural investment], *An Giang*, 23 October 1987, p. 2; Author's interviews, June–August 2005, Long Điền B.
73   Author's interview, 6 August 2005, Long Điền B.

I could not count them all. Some did not pay a thing for four successive seasons. Most debtors were land recipients who previously had not had any land [*không có cục đất chọi chim*]. After receiving land, they did not transform or level out the land. They did not know how to farm. So when they grew rice on land that was soaked here but dry there, some of their rice died, some survived. The results of such farming were bad, so they owed the production unit. Meanwhile, professional cultivators knew how to farm and had capital to invest. They levelled out land properly so their crops grew better.[74]

A former cadre of Long Điền B commune observed that, among debtors, the 'priority' families (*gia đình chính sách*)—such as those with wounded veterans or who had members killed fighting in the war (*thương binh liệt sỹ*), former soldiers and local poor cadres—had the largest debts. Apart from farming unproductively, these households often took advantage of their position to evade paying the production units.[75] To collect arrears and make peasants pay their debts, production unit cadres in Long Điền B threshed peasants' paddy, especially those who were in debt or had poor harvests. A leader of production unit no. 15 recalled:

The production unit had to control produce. After harvesting and threshing paddy in the fields, each household had to pay [fees, taxes and funds] before carrying paddy home. If a production unit allowed individual households to harvest freely, they would refuse to pay their debts fully. It was common that households with low paddy productivity came to reap and hide paddy at night. So, during the harvest time, production unit cadres had to patrol the fields at night.[76]

Villagers in Long Điền B recalled that, apart from collecting fees in the fields, production unit cadres and commune police frequently searched debtors' houses and confiscated their paddy and/or belongings; they even arrested some and held them in custody. However, the results were insignificant. Many debtors continued to refuse to pay their debts, justifying their behaviour by saying they had no means to pay.[77] The *An Giang* newspaper was critical, reporting in July 1988:

---

74 Author's interview, 16 August 2005, Long Điền B.
75 Author's interview, 29 July 2005, Long Điền B. The households of martyrs and wounded soldiers included those who had family members killed or wounded in the line of duty during the country's wars. In An Giang, most of these were related to conflicts with Cambodia (1978 to mid-1980) and China (late 1970).
76 Author's interview, 16 August 2005, Long Điền B.
77 Author's interviews, June–August 2005, Long Điền B.

Over the past years, peasants have not had the right to manage and control their own land and produce. Their fate was determined by others [cadres]. The only right that they had was to labour … It was common that local authorities came to search for peasants' paddy, confiscated their belongings and took back the land to reduce their debts. It was a daily phenomenon that peasants in debt were arrested and remained behind bars for so-called 'education'. Many peasants did not have enough food; how could they pay?[78]

After *Đổi Mới* started, villagers tended to refuse to fulfil obligations such as contributing collective funds to production units. A former cadre of production unit no. 16 recalled that, in the late 1980s, the number of peasant households which refused to pay production unit funds increased in Chợ Mới district. These households were fed up with unprofitable farming in production units and discontent with cadres' embezzlement and incompetence. In response, local authorities used harsh measures, such as sending soldiers to search for paddy and belongings in their homes. They even arrested debtors. However, these tactics did not result in significant changes, and dismayed peasants. The cadre concluded, 'when peasants refused to contribute to funds, the only option was to dismantle the production units'.[79]

Abandoning, transferring, mortgaging and disputing ownership of land

Instead of cultivating redistributed land, some poor peasants in An Giang transferred and mortgaged it to others because they did not farm or did not have enough capital.[80] A former chairman of the farmers' association in Long Điền B commune said that, after a few seasons of farming, some peasants abandoned, transferred or mortgaged their land so they could take up non-farming work.[81] Another man in Long Điền B said it was common for land recipients to transfer or sell their land to others. Some fields had been transferred several times between different landholders.[82]

---

78   Ý kiến: Làm chủ [The opinion piece: Being a master], *An Giang*, 29 July 1988, p. 1.
79   Author's interview, 11 July 2005, Long Điền B.
80   Võ Tòng Xuân and Chu Hữu Quý, *KX Account 08-11*, p. 35.
81   Author's interview, 9 August 2005, Long Điền B.
82   Author's interview, 2 August 2005, Long Điền B.

Villagers and former cadres in Long Điền B argued that transferring land made those who had previously had their land redistributed discontent with the redistribution policy. Moreover, land-givers became more discontented when they saw that some of the local cadres took advantage of their positions and misappropriated the land for themselves and their relatives.[83] However, before the *Đổi Mới* policy, many landowners did not express their views publicly.

In the spirit of *Đổi Mới* and in response to the poor performance of agriculture, on 19 February 1987, An Giang people's committee issued Decision No. 93-UBND, which corrected the mistakes of previous land redistribution efforts. It also allowed households to farm fields in communes other than their own.[84] It advocated retrieving land that had been previously appropriated 'irrationally and illegally' by local cadres and state organisations and giving it back to the former landowners according to their capacity to farm or to those who were currently landless or land-poor. The policy triggered a host of claims from households for their former land and led to widespread conflict among peasants and local cadres in rural areas of An Giang.[85]

According to the *An Giang* newspaper, within the first three months of 1987, authorities in the province had received 2,000 letters from peasants and met 5,000 people who submitted petitions. Most letters complained about land, houses, belongings and agricultural and other machines being appropriated or transformed (*cải tạo*) in ways that violated people's mastery rights. Some letters accused local cadres of embezzling collective resources.[86] Further, in May 1987, *An Giang* reported that, during the implementation of provincial Decision No. 93, a complicated problem had emerged when many former landowners claimed back their land from the new owners. This problem occurred almost everywhere. Regardless of the local authority's decision, some

---

83 Author's interview, 16 August 2005, Long Điền B.

84 Võ Tòng Xuân and Chu Hữu Quý, *KX Account 08-11*, p. 43; Hội nghị cán bộ quán triệt nghị quyết 1987 [Officials' meeting thoroughly resolves 1987 resolution], *An Giang*, 22 December 1986, p. 1.

85 Cần hiểu rõ và chấp hành tinh thần quyết định 93 của Ủy ban nhân dân tỉnh [The need to fully understand and abide by the Provincial People's Committee's Directive No. 93], *An Giang*, 22 May 1987, p. 6; Don't mistake rational reallocation, *An Giang*, 29 May 1987, p. 1.

86 Phỏng vấn phó bí thư tỉnh ủy An Giang: Nhiệm kỳ tới sẽ cố gắng làm thế nào để góp phần vận dụng nghị quyết VI vào thực tế tỉnh nhà đạt kết quả cụ thể hơn nữa [Interview with Vice-Chairman of An Giang's Party Committee: The next term will try to contribute to the application of Resolution VI in the province to achieve more concrete results], *An Giang*, 17 March 1987, p. 1.

peasants claimed their former land by illegally sowing seeds on it or by other means.[87] For example, in 1985, the production unit cadres in Đức Bình ward, Long Xuyên town, redistributed 28 *công* of surplus land from a Mrs Kiem to seven other households to establish product contracts. In April 1987, however, Mrs Kiem planted seeds on these 28 *công*, in defiance of previous arrangements. In the end, authorities had to force her to return the land to the seven new land users (*chủ mới*). Mrs Kiem was not alone; 14 other landowners tried to take back their old land in this area.[88]

To tackle these problems—of previous landowners (*chủ cũ*) trying to reclaim their redistributed land—authorities issued several announcements stressing 'the need to understand clearly and conform to the spirit of Decision No. 93' and accusing former landowners of mistaking (*ngộ nhận*) the policy of 'reallocating land rationally' for one of 'returning land to previous owners'.[89]

Land disputes were widespread not only in An Giang, but also across the Southern Region. According to researcher Huỳnh Thị Gấm, by August 1988, 59,505 peasants had lodged complaints about land across the whole Southern Region. In many areas, peasants took back their former land or fought against each other, state enterprises and military organisations. There were physical clashes and incidents in which people were wounded and some were killed. For example, seven people died in An Giang and Cửu Long in 1988. Peasants also organised demonstrations. They gathered together, carrying national flags, slogans and pictures of former national chairman Hồ Chí Minh. They marched through government offices at all levels, from the commune, district and provincial to the central, demanding resolution of their land claims.[90]

---

87 Mỗi tuần một chuyện: Nhanh chóng giải quyết vấn đề ruộng đất hợp lý [A story each week: Be quick to solve land disputes], *An Giang*, 22 May 1987, p. 3.
88 Mỗi tuần một chuyện: Hiểu lầm hay cố ý? [A story each week: Misunderstood or intended?], *An Giang*, 29 May 1987, p. 7.
89 The need to fully understand and abide by the Provincial People's Committee's Directive No. 93, *An Giang*, 22 May 1987, p. 6; Don't mistake rational reallocation of land, *An Giang*, 29 May 1987, p. 1.
90 Huỳnh Thị Gấm, Socioeconomic changes in the Mekong Delta, p. 89. According to Huỳnh Thị Gấm, by the end of 1988, authorities in An Giang had received 41,000 petition letters from peasants, Đồng Tháp had received 20,000, Minh Hải 18,000 and Cửu Long 10,000.

In response to these disputes, the VCP issued Directive No. 47/CT-TW (31 August 1988), which recognised the shortcomings of previous land redistributions. First, land redistribution, especially under Directive No. 19 (3 May 1983), had equalised (*cào bằng*) landholdings among households and interrupted and rearranged previous farming systems (*xáo canh*) in rural areas, which negatively affected agricultural production. Second, it distributed land to non-farming households such as small traders and other non-farm workers who did not know how to farm. Finally, local cadres and state agencies had taken advantage of the policy to use land inappropriately. Now peasants wanted their old land back.[91]

To boost commodity production, the directive advocated the elimination of the prohibition on non-resident cultivators and the retrieval of land that was farmed poorly or illegally. Retrieved land was to be redistributed to productive landowners or to those who currently had insufficient land. However, the directive called for land disputes to be dealt with cautiously, case by case. It also stipulated that landlords, reactionaries, rich peasants and rural capitalists whose land had been confiscated under the policy to eliminate exploitation did not have the right to reclaim their land.[92]

To clarify the central government's Directive No. 47, leaders in An Giang issued Directive No. 303-QD-UB (4 October 1988)—which contained a feature not mentioned in the central directive. It encouraged former landowners and the new users of land to negotiate with each other to determine who should own the land and to decide on the level of any compensation. For example, if a new land user wanted to keep the land, they would have to compensate the former landholders for the cost of land reclamation and transformation (*công khai phá và cải tạo ruộng đất*), although this would be applicable only to middle peasant households. Otherwise, the former landowner must compensate the new land user for the cost of rehabilitating and transforming the land and the value of the crop on it.[93]

---

91 Ban Tuyên Huấn Trung Ương, *The Party's Response to Urgent Land Problems*, p. 9.
92 Ibid., pp. 9–14.
93 UBNDTAG (1989), *Quyết định 303/QĐ-UB* [*An Giang People's Committee Directive No. 303*], 4 October, Long Xuyên: An Giang.

Directive No. 303 triggered a second wave of land disputes in An Giang. One popular rumour was that authorities would return land to its former landowners. Excited by this news, many former landowners in Long Điền B rushed to claim their land. Some met new land users to negotiate the return of their land; some simply brought seeds to sow on their old land regardless of what the authorities said; and some gathered at commune and district offices to strike and demand resolution of their land claims. All of this caused what villagers in Long Điền B called 'great turmoil' (đảo lộn) in Chợ Mới district and elsewhere in An Giang during the late 1980s.[94] A former vice-chairman of Long Điền B commune shared a story of how one former landowner whose family had two ploughing machines responded to Directive No. 303:

> Hearing news of Directive No. 303, before anyone was ready to work the fields, he and his brothers carried long knives and machines to plough his family's old land. They threatened to kill anyone who dared to block them. So, the new land users did not dare to. Finally, the commune police had to arrest them. At the office, they argued that authorities had redistributed their land to others to do collective farming in production units, but now collective farming in production units did not really exist any more so the authorities had to return the land to them.[95]

A Long Điền B resident who had lost 6 hectares of land in Long Điền A commune due to the non-resident cultivator prohibition, and who later took that land back, recalled:

> Before the reunification, I had 6 hectares of land [in Long Điền A]. After reunification, revolutionary authorities took all my land to redistribute to others. They took my land right out of my hands. The hamlet chief in Long Điền A appropriated much of my land. However, after Mr Linh came to power, I had a chance to take it back. I also sent many letters to claim my land, but authorities rejected them all. So, I decided to break the law. My two brothers and I brought machetes to the field to work; I said that if he [the hamlet chief] came to the field, we would kill him. I said that it was right for the authorities to take abandoned land, but not right to steal land from people. Thanks to the party secretary of Long Điền A commune, who asked the hamlet chief to return the land to me, I was able to get the land back.[96]

---

94 Author's interviews, June–August 2005, Long Điền B.
95 Author's interview, 29 August 2005, Long Điền B.
96 Author's interview, 9 August 2005, Long Điền B.

A former cadre of production unit no. 1 who was aware of land conflicts after Directive No. 303 recalled:

> Directive No. 303 did not tell the new users of land to return land to the old landowners. It just mentioned that both needed to negotiate with each other in the spirit of mutual concession. But it seemed that the authorities favoured the interests of the old landowners. I did not know what provincial leaders' opinions were, but I knew that some district and commune cadres implicitly supported returning land to the old landowners [to boost commodity production]. As far as I remember, at the meeting to deal with land disputes in 1988, Mr Chau, a district leader, said that people could not get rich with 2 and 3 *công* of land. With a few of *công* of land, people could not produce commodity paddy. So, people should return land to old landowners and find other businesses. Therefore, in land disputes, old landowners had the advantage over new land users. Eventually, most new land users in Long Điền B decided to return land to old landowners.[97]

As discussed in previous chapters, during land redistribution, authorities in Long Điền B allowed landowners to 'lend' (*cho mượn*) much of their surplus land to their land-poor relatives. Moreover, Long Điền B villagers highly respected the rights of individual landownership and values of justice and religious morality. Therefore, new land users tended to return land to its former owners. A man who lost his land in Đồng Lớn later received 3 *công* of land from his relative, but, after Directive No. 303, he felt emotional about his relative (*tình cảm bà con*) so he decided to return the land. He did not want to fight over it because it would bring him a bad reputation (*mang tiếng*).[98] Similarly, a landless man who received a few *công* from an acquaintance decided to return it to its previous owner. He felt it was odd for him to take another person's land, and he would rather be poor than steal someone else's land (*giụt đất người khác*).[99]

Some landowners in Long Điền B still complained they were unable to reclaim their land, especially land located in other communes, districts or provinces. Notably, peasants who had land in Đồng Tháp said they could not get it back as the authorities there favoured their own residents. However, landowners who had lost fields in Thoại Sơn and Châu Thành districts in An Giang province were able to take back much of their

---

97   Author's interview, 17 August 2005, Long Điền B.
98   Author's interview, 3 August 2005, Long Điền B.
99   Author's interview, 16 August 2005, Long Điền B.

land during the late 1980s and early 1990s. In general, from 1988 to the early 1990s, many upper–middle and middle peasants retrieved much of their previous landholdings. An upper–middle peasant who had lost 200 *công* of land in Thoại Sơn recalled: 'Thanks to Mr Linh, I could retrieve half of my land and a ploughing machine. I was very happy when I took it back. People should worship Mr Linh!'[100]

According to one report, from 1988 to 1990, An Giang had dealt with more than 30,000 peasant complaints about land, which reduced tensions in rural areas.[101] However, the legacy of collectivisation and land disputes in An Giang was not solved yet. Still today land disputes are a hot issue in rural areas of An Giang and elsewhere in the Southern Region.

Local cadres' practices in An Giang

Under the product contract system, collectivisation in An Giang accelerated. This required additional local cadres and increased efforts to supervise them. Yet, despite the efforts of An Giang's leaders to improve the quality of local cadres during the period 1981–88, problems remained and even seemed to worsen. The most common malpractices among local cadres were misappropriation of land and embezzlement of state agricultural inputs (fertilisers, pesticides, fuel and so on).

Misappropriating collective inputs and funds

An Giang newspaper accounts highlighted numerous cases of local cadres who exploited their positions to misappropriate and embezzle agricultural inputs. For example, a report on 27 June 1982 said that, after receiving state provisions of fertiliser and fuel, a commune cadre in Châu Thành district sold them on the 'black market' for a quick profit rather than giving them to peasants, according to the requirements of the B contracts. The commune cadre used his ill-gotten gains to upgrade his house and pigpen and reported to the higher-level authorities that peasants had refused to pay their input debts.[102]

---

100 Author's interview, 18 August 2005, Long Điền B.
101 UBNDTAG, *An Giang Province*, p. 400.
102 Chuyện to nhỏ: Ông cán bộ xã T [Some issues: Commune cadre], *An Giang*, 27 June 1982, p. 4.

In March 1983, *An Giang* reported:

> Recently, some peasants complained that local cadres were stealing production unit inputs to sell on the black market for a quick profit or to raise [black] funds. Some cadres even misappropriated inputs worth 15 tonnes of paddy ... This made agricultural production in An Giang difficult.[103]

In 1984, An Giang put several local cadres on trial for embezzling agricultural inputs, collective paddy and other goods. In July 1984, Tran Van Ba, a Long Xuyên agricultural input station accountant, was put on trial for colluding with leaders of production units and production solidarity teams and misappropriating a large quantity of inputs to sell on the black market.[104] In September, an accountant from Bình Long commune's food station was taken to court for writing fake invoices and embezzling 3,027 kilograms of state paddy to sell on the black market.[105] In December, the An Giang people's court tried 26 cadres who were staff of the provincial food department. They were accused of increasing the price of cement, which the Ministry of Food used to exchange for paddy with peasants. In addition, they created fake receipts for millions of dong, embezzled, took bribes and stole state inputs. In the end, the court sentenced one of them to death; the head of the provincial Ministry of Food was sentenced to 17 years' imprisonment and others were sentenced to many years in jail.[106]

Despite continuous efforts by the An Giang authorities to tackle cadres' misbehaviour, problems persisted. On 16 May 1986, *An Giang* reported that two-thirds of provincial goods used to exchange for peasants' paddy had fallen into the hands of individual merchants, most of whom were relatives of local cadres.[107] Local cadres unilaterally increased the prices of goods and agricultural inputs that peasants had to purchase from state agencies. The cadres then sold considerable amounts of these goods and inputs on the black market, making quick profits. This contributed

---

103 Chuyện to nhỏ: Nên chấm dứt [Some issues: Stop it], *An Giang*, 13 March 1983, p. 4.

104 Tòa án nhân dân tỉnh xét xử đầu cơ và hối lộ [Provincial People's Court adjudication on speculation and bribery], *An Giang*, 12 July 1984, p. 3.

105 Huyện Châu Phú xét xử bọn tham ô lương thực [Châu Thành District Court tries food thieves], *An Giang*, 27 September 1984, p. 3.

106 Phạm nhiều tội, 26 bị cáo ra tòa án nhân dân An Giang [26 defendants sentenced for many crimes in An Giang People's Court], *An Giang*, 27 December 1984, p. 2.

107 Hàng đổi hàng đến tay ai? [Who benefits from goods exchanged for paddy?], *An Giang*, 16 May 1986, pp. 3–4.

to inflation in the province and aided the survival of black markets and individual merchants—something the VCP leadership was trying to control and eliminate.[108]

At the local level, production unit cadres played an intermediary role in economic transactions between the state and peasants, so they had even more opportunities to capture resources. It was common for villagers to send letters to newspapers or the state to accuse cadres of embezzling agricultural inputs and collective property. For example, in 1984, peasants in production unit no. 12 in Kiến Thành commune (Chợ Mới district) sent a letter accusing their production unit cadres of buying things without receipts. They also accused them of stealing agricultural inputs and overcharging members for the costs of production (such as fuel, fertilisers and collective funds).[109]

Likewise, according to the *An Giang* newspaper in August 1984, peasants in one production unit were surprised to see the paddy fields of production unit cadres were more luxuriant and had higher productivity than those of ordinary peasants, when all were supposed to be operating under the same conditions. An inspection found that production unit cadres had taken scarce agricultural inputs, such as fertilisers, pesticides and fuels, for themselves, rather than distributing them among households. They also sold some of these products on the black market. This explained why, after a few years of working as cadres, all had newly renovated houses and expensive belongings.[110]

Villagers in Long Điền B asserted that production unit cadres embezzled a considerable sum of agricultural inputs and collective funds. From 1982 to 1987, inspections discovered many production unit cadres embezzling agricultural inputs and collective paddy. Some were imprisoned—for example, a former production unit head was sentenced to several months in prison in 1986 for embezzling 75 *giạ* of paddy (according to him), but 16 tonnes of paddy according to his successor.[111] A man in production unit no. 9 recalled:

108 Chuyện to nhỏ: Xé rào [Some issues: Fence breaking], *An Giang*, 31 October 1982, p. 4; Xã luận: Tăng cường quản lý thị trường và ổn định giá cả [The editorial: Strengthening control of markets and stabilising prices], *An Giang*, 22 May 1983, p. 1; Who benefits from goods exchanged for paddy?, *An Giang*, 16 May 1986, pp. 3–4.
109 Chuyện to nhỏ: Đề nghị giải quyết thỏa đáng [Some issues: The need to solve the problem satisfactorily], *An Giang*, 28 August 1983, p. 4.
110 Chuyện to nhỏ: Chuyện các ngài trong ban quản lý tập đoàn [Some issues: The problems caused by production unit managerial cadres], *An Giang*, 8 October 1984, p. 4.
111 Author's interviews, 9 and 11 August 2005, Long Điền B.

The production unit cadres served people very poorly, but embezzled very well. Their pockets were full from embezzlement. For example, when pumping water to peasant fields, it cost one container of fuel, but they reported three. When raising funds to buy farm machines, instead of charging each production unit member 30 kilograms of paddy per 1 *công* of their land, they charged 34 kilograms. So, how much would they get for about 1,000 *công* of land?[112]

The inspections in Chợ Mới during the second quarter of 1986 also showed violations in several managerial boards of production units. For example, in Bình Hòa commune, a storekeeper for production unit no. 17 embezzled 3,235 kilograms of paddy; a storekeeper for production unit no. 15 embezzled 6,051 kilograms; and a leader of production unit no. 7 misappropriated 6,244 kilograms of collective grain.[113] A former cadre of the Chợ Mới Committee for Agricultural Transformation recalled:

Most production units were not low in quality; they were production units on paper and ghost units [*tập đoàn giấy, tập đoàn ma*]. When inspecting, we discovered violations in many of them. Because the inspection was to strengthen production units, we did not take them to court [*đưa ra pháp luật*]. For example, in 1986, the provincial inspection in Mỹ Hội Đồng and Mỹ Lương communes uncovered many cases of cadre embezzlement, but they were settled internally [*xử lý nội bộ*], not in public.[114]

In explaining why so many local cadres embezzled, a primary schoolteacher from Long Điền B commented that production unit leaders were selected from revolutionary and pro-revolutionary families; most were poorly educated but were recruited because they readily accepted the posts (while better educated people were reluctant to take on such positions). They had also not been well trained, so they managed the units ambiguously and poorly. The teacher concluded: 'All cadres in production units and interproduction units embezzled collective resources.'[115]

---

112 Author's interview, 10 August 2005, Long Điền B.
113 Huyện Chợ Mới tiến hành kiểm tra một số tập đoàn nông nghiệp [Chợ Mới district carried out investigations into some production units], *An Giang*, 13 June 1986, p. 1.
114 Author's interview, 17 June 2005, Chợ Mới.
115 Author's interview, 15 August 2005, Long Điền B.

Cadres' debts

An Giang newspaper accounts suggest cadres' improper behaviour continued and even worsened. In November 1984, the provincial food department found that peasants' debt was small compared with that of local cadres. For example, four communes in Châu Phú district each owed 400 to 500 tonnes of paddy; most debtors were local cadres.[116] In May 1985, many areas of Châu Thành district still had huge debts; some communes owed 600 tonnes of paddy each, according to a provincial Inspection Commission report. And most of the large debtors were cadres. For example, in A. H. commune, the chairman owed 80 tonnes of paddy; the chief and the storekeeper of the commune's Department of Agricultural Inputs owed 14 and 16 tonnes of paddy, respectively.[117] In Bình Hòa commune, 30 of 36 production units were in debt, much of it due to cadres' theft. Fearing punishment, some production unit cadres ran away (*bỏ trốn*).

Commune and district cadres in Châu Thành also had large debts. For example, Duong Van Minh, a district irrigation agent, owed 4 tonnes of paddy, while Vo Van Rang, a district inspection agent, owed 806 kilograms. Such officials, according to a 1987 report, 'took advantage of their positions and the weakness of loose management to collude and steal state agricultural inputs. Some cadres owed 50–70 tonnes, even more.'[118]

Similar to Châu Thành district was Thoại Sơn. From 1983 to the winter-spring of 1986–87, that district had 21,500 tonnes of paddy debt, of which input (B contract) debt was 15,000 tonnes and unpaid taxes were 6,500 tonnes. According to a manager of the district's food company, the debts of production unit, commune and district cadres accounted for 70 per cent of the total; ordinary peasants' debt was only 30 per cent. Moreover, despite commune and district cadres' families owing large debts, production unit cadres did not dare collect because they 'feared higher officials' (*tâm lý sợ cấp trên*).[119]

---

116 Chuyện to nhỏ: Nợ không chịu trả [Some issues: Refusing to pay outstanding debt], *An Giang*, 16 November 1984, p. 4.
117 Cited from Chuyện to nhỏ [Some big and small issues], *An Giang*, 24 May 1985, p. 4.
118 Tại sao Châu Thành chưa giải quyết được tình trạng nợ trầm trọng? [Why haven't Châu Thành district authorities dealt with their huge outstanding debt?], *An Giang*, 31 May 1985, p. 3.
119 Tình hình thanh lý nợ hợp đồng trong sản xuất nông nghiệp ở Thoại Sơn [Contract debt liquidation in agricultural production in Thoại Sơn district], *An Giang*, 4 September 1987, p. 3.

Phú Tân district experienced similar circumstances. From the winter-spring of 1986–87 to June 1987, the total debt of 20 party members and 50 production unit cadres reached thousands of tonnes of paddy. Some owed 40–50 tonnes of paddy each.[120] In Hòa Lạc commune (Phú Tân), 24 of 27 commune party cell members owed more than 1 tonne of paddy each.[121]

Local cadres' debts were large and common in many parts of An Giang. The provincial newspaper in July 1987 said commune, hamlet and production unit cadres in the province owed about 70 per cent of the total B contract debt.[122]

In late 1987, provincial leaders decided to revise the policy on agricultural inputs. Local cadres' poor management and embezzlement hindered the accurate and timely delivery of inputs, causing difficulties for production and peasants' livelihoods. From the winter–spring of 1987–88 onwards, An Giang decided to end the delivery of state agricultural inputs to peasants through production units. Instead, state inputs would be sold to peasants directly in exchange for cash or paddy.[123]

Misappropriating peasants' land

Cadres in many parts of An Giang were accused of misappropriating (*chiếm dụng*) peasants' land, accounts of which emerged after *Đổi Mới*, and especially after the provincial people's committee issued its Decision No. 93 (19 February 1987). In May 1987, the *An Giang* newspaper pointed out:

> Over the past years, the redistribution of provincial land was irrational. Some cadres have taken advantage of their position to gain good land for themselves and their families. Others did not cultivate the land but took a considerable portion of it. Many state organisations at district and provincial levels took advantage of their collective status to misappropriate land.[124]

---

120 Cai hợp đồng B [Boss of the B contracts], *An Giang*, 28 August 1987, p. 7.
121 The problem of agricultural investment, *An Giang*, 17 July 1987, p. 2.
122 Ibid.
123 Peasants need a rational method of agricultural investment, *An Giang*, 23 October 1987, p. 2.
124 Don't mistake rational reallocation, *An Giang*, 29 May 1987, p. 1.

In July 1988, the *An Giang* newspaper listed the names of several cadres who had used the prohibition against non-resident cultivators to take land for themselves. For example, Cao Hồng Dinh, Tân Lập commune police chief, whose family already had 2 hectares of land, took another 6 hectares; Ba Hương, the commune's Department of Agriculture head, appropriated over 10 hectares; and Tứ Dũng, the vice-commune chairman, took more than 12 hectares. Some peasants whose fields had been usurped 'lost their temper' (*loạn trí*) and went to commune offices, shouting and demanding their land back.[125]

Several government offices and mass organisations took land for illicit purposes. In Định Mỹ commune in Thoại Sơn district, for instance, such organisations took more than 160 hectares. They tried to justify this by calling the areas 'self-sufficient land' (*đất tự túc*) to serve the benefit of the entire organisation. Although annoyed, villagers initially tolerated this behaviour, but certain officials ended up using the land as their own. For example, the commune's party secretary took 6 hectares of land for himself; the party's vice-secretary took 12 hectares; the commune's vice-chairman took 12 hectares; and the chief of the commune police took 6 hectares. One group of cadres claimed a vast 167 hectares, which they classified as unclaimed land (*đất hở*). A Mr Cop, a cadre of the commune's Department of Agricultural Tax, took (*bao chiếm*) 26 hectares for himself, an act he hid by using the names of seven different landholders. Lê Văn Dũng, the chief of the same department, appropriated 14 hectares and hid the theft under four different names. Mr Tân, the chief of communal police, stole 31 hectares, while Út Hên, the commune vice-chairman, took 31 hectares, using the names of different landholders.[126]

Land misuse was also severe in Phú Tân district. In 1982, district authorities prohibited non-residents from farming there. Taking advantage of the situation, many local cadres took fields for themselves. For example, Trần Văn Phát, the leader of production unit no. 17 in Long Phú commune, took more than 2.7 hectares; Nguyễn Văn Hảo, the leader of production unit no. 15, took 5.4 hectares; Tô Văn Ba, the chairman of Long Phú commune's Father Front (a social organisation), took 1 hectare; Út Bình, the former commune chairman, appropriated

---

125 Nỗi oan trái của bà con nông dân Tân Lập [The grievances of peasants in Tân Lập], *An Giang*, 1 July and 8 July 1988, p. 3.
126 Những người bao chiếm đất [Land misusers], *An Giang*, 5 August and 19 August 1988, p. 2.

5 hectares; Chau Ngoc Chao, the commune chairman, appropriated 5 hectares; and Nguyễn Văn Thái, the commune's party secretary, appropriated 4.5 hectares. By August 1988, hamlet and production unit cadres had taken 78 hectares from non-resident farmers.[127]

Cadres also misused land in Chợ Mới. A man in Long Điền B admitted that local cadres misappropriated land everywhere, and it was common for cadres to have more land than ordinary people.[128] Another man said compared with land appropriation elsewhere, in Chợ Mới it was less severe.[129] However, in some Chợ Mới communes, according to accounts in the *An Giang* newspaper, land appropriation was just as bad as elsewhere. For example, in Tân Mỹ commune in Chợ Mới district, many cadres took peasant land, concealing their action under different names. Some even resold the land to make a quick profit. Cadres delayed or, in the worst cases, avoided implementing state policy on returning land to the previous landowners. So, nearly two years after Decision No. 93 (19 February 1987) had been issued, authorities in Tân Mỹ commune had not settled any peasants' land claims.[130]

Other bad practices

An Giang villagers often accused cadres of monopolising and overcharging for farming services to production units. As discussed earlier in this chapter, households in most production units in An Giang did almost all phases of farmwork themselves.[131] Production unit cadres, however, controlled certain farming resources and services, such as irrigation and equipment for ploughing, raking and threshing. Often cadres and the specialised teams responsible for providing or using these resources were inefficient or unfair in how they provided services. A man in Long Điền B recalled how irrigation was done in his fields:

---

127 Phú Long: Cán bộ xã còn bao chiếm đất [Phú Long: Cadres still misappropriate land], *An Giang*, 9 December 1988, p. 4.
128 Author's interview, 29 August 2005, Long Điền B.
129 Author's interview, 27 July 2005, Long Điền B.
130 Đất: Tiếng kêu từ phía nông dân [Land problem: A cry from peasants], *An Giang*, 18 November 1988, p. 3.
131 Chuyện to nhỏ: Ông tập đoàn trưởng [Some issues: Production unit leader], *An Giang*, 13 June 1982, p. 4.

Production unit, my goodness! Production unit members had to compete with each other to have their land watered. We had to draw lots to determine who was served first. If we were first, we had to spend days and nights guarding the water. Within two days, if we hadn't finished watering, we had to give the water to others and waited for another turn.[132]

A woman from Long Điền B added: 'The production unit teams irrigated for some people and not for others. When irrigating fields, some places got too much water, others nothing.'[133] A man in production unit no. 9 in Long Điền B remembered problems getting his fields ploughed:

We contributed paddy to the production unit to buy ploughing and threshing machines but we still had to pay for ploughing and threshing. They were not free of charge. Moreover, the guys controlling the machinery served their relatives first rather than the rest of us. In order to have our land ploughed, we had to entreat [năn nỉ] them five or 10 times and always carry cash to pay them right away. Otherwise, they would not plough our land.[134]

In April 1983, the *An Giang* newspaper reported that, in a certain district, only a few collective ploughing machines operated even at peak times of land preparation. Moreover, their ploughing capacity was extremely low. The reason, according to the article, was that operators of the collective machines were waiting for 'special fuel' (bribes) from peasants, which was 'necessary for machines to run fast'.[135]

In July 1986, the newspaper reported that peasants in one interproduction unit had criticised cadres for poor ploughing services:

When ploughing, equipment operators just ploughed around the plot, leaving the centre untouched … [and] often the tractors ran like a racehorse [chạy như ngựa đua] and raked like a mouse scratching the land [xới như chuột cào].[136]

---

132 Author's interview, 2 August 2005, Long Điền B.
133 Author's interview, 12 August 2005, Long Điền B.
134 Author's interview, 5 August 2005, Long Điền B.
135 Lệ Làng [Village customs], *An Giang*, 24 April 1983, p. 4.
136 Tự phê bình và phê bình: Ý kiến từ một cuộc họp [Criticism and self-criticism: Opinion from a meeting], *An Giang*, 4 July 1986, p. 3.

As a result, the fields were poorly prepared. In addition, the interproduction unit cadres often rented out tractors to other areas instead of fulfilling their obligations to members of their own interproduction unit.[137]

A former cadre of Long Điền B commune complained about the performance of disease prevention teams (*đội bảo vệ thực vật*):

> During that time [of production units], peasants complained a lot about these teams because they performed very poorly. They called the crop protection team *đội bảo vệ thịt vịt* [the 'duck meat protection team'] because only by giving the teams duck meat did they work well. Otherwise, they worked badly. In the end, we let peasants receive pesticides and spray their own crops.[138]

One of the most annoying things for villagers in Long Điền B was that production unit cadres monopolised the threshing service for peasants' paddy. One man recalled:

> The production unit took over threshing our paddy without allowing others [other production units or individuals] to do the job, regardless of whether it was raining or not. They also overcharged us.[139]

Similarly, another man recalled the way his interproduction unit threshed:

> The interproduction unit [including four production units] had four threshing machines, so each production unit had one machine. How could they thresh people's paddy in time? They did not allow people to hire outside threshing services. When it rained, people's paddy got wet and rotted. Seeing their paddy going to ruin, some people got so angry that they lay down on the road where cadres passed their threshing machines and shouted, 'Thresh my paddy or kill me!'[140]

Local cadres' malpractice contributed to the derailment of national and provincial agrarian policies. By capturing such a large proportion of state and collective resources and serving farmers poorly, local cadres exacerbated the poor performance of collective organisations, and of agriculture in the south as a whole. While one of the original aims

---

137 Ibid., p. 3.
138 Author's interview, 29 August 2005, Long Điền B.
139 Author's interview, 27 July 2005, Long Điền B.
140 Author's interview, 16 August 2005, Long Điền B.

of the VCP's agrarian reforms was to eliminate exploiting classes, in many areas, rural cadres became a new class of exploiters. The *An Giang* newspaper reflected critically that, 12 years after the country's reunification, peasants should have escaped poverty and backwardness. However, having just escaped from the darkness of landlordism, peasants were exploited by 'new landlords' (*địa chủ mới*) masked in the name of production units.[141] The following cartoons (Figures 7.1–7.4) from the *An Giang* newspaper depict cadres' other misdemeanours in An Giang during the period of product contracts.

Quỹ nội bộ . . .

Tranh : Nguyễn Ngô

Figure 7.1 Internal funds

Behind the accountant of a production unit were several types of 'internal funds' [*quỹ nội bộ*] used only by cadres themselves.

Source: Drawn by Nguyen Ngo, published in *An Giang*, 2 August 1983, p. 4.

---

141 Những điều nghe thấy từ thực tế [Some issues learned from reality], *An Giang*, 4 March 1988, p. 3.

Figure 7.2 Red tape

A peasant who submits his petition to the boss in a state office must go through several gates. The first gatekeeper asks, 'Do you have permission papers?' The second gatekeeper asks, 'Do you want to meet the boss? Wait here.' After considering the form, the secretary replies: 'Approved, come and pick up the results in a few days.' But the cadre behind the secretary says: 'Finished, come and pick up the result in a few months.' Flooded with piles of petition letters, the boss shouts: 'Go back home! I will sign later after studying it.' Finally, the peasant wonders: 'But we are told that red tape has been eliminated!'

Source: Drawn by T. Q. Vu, published in *An Giang*, 2 October 1987, p. 7.

Figure 7.3 Prohibiting the use of cameras

While preparing a lavish party, the boss orders a staff member to post a big sign prohibiting the use of cameras so that, the boss says, 'We need not fear being photographed by journalists!'

Source: Drawn by T. Q. Vu, published in *An Giang*, 2 September 1987, p. 7.

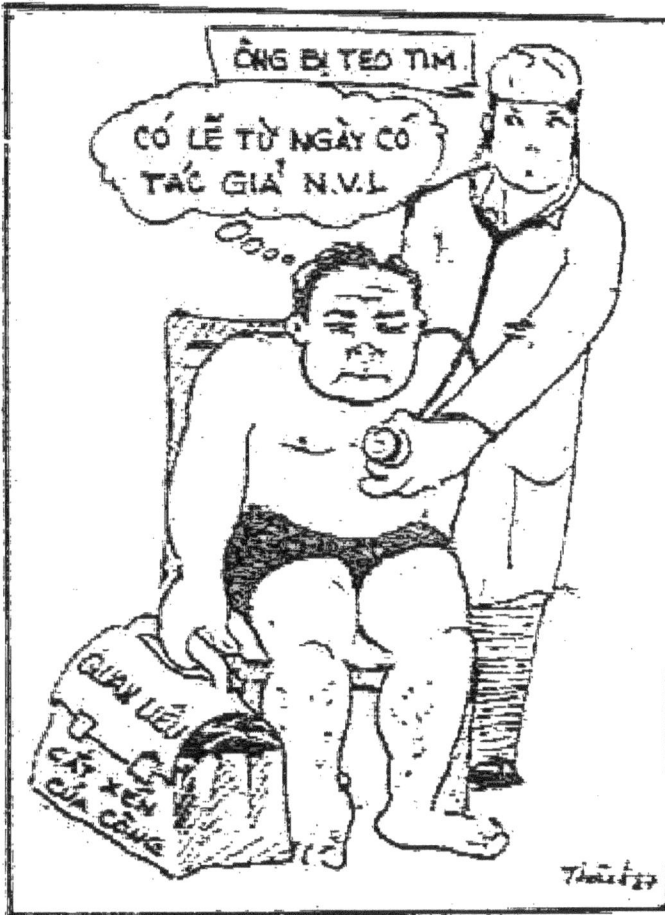

Figure 7.4 Heart problem

After examining a cadre who has benefited from bureaucratic red tape and embezzlement, the doctor says: 'You have a heart problem!' The cadre ponders: 'Probably I have had this problem since the appearance of NVL.'[142]

Source: Drawn by Van Thanh, published in *An Giang*, 4 September 1987, p. 7.

---

142 NVL is the abbreviation of the name of the Communist Party General Secretary, Nguyễn Văn Linh, who initiated the *Đổi Mới* policy and cracked down on corruption. 'NVL' was often also interpreted as '*Nói và Làm*' ('speaking and doing').

A return to household farming

QN-ĐN in the Central Coast

According to QN-ĐN government statistics, in the first few years of the product contract system, staple food production in the province increased from 460,000 tonnes in 1980 to 500,000 tonnes in 1981 and 525,000 tonnes in 1982.[143] However, from 1983 to 1985, staple food production stagnated, and then decreased to 510,000 tonnes in 1983.[144] It increased slightly in 1984 (to 522,000 tonnes) and in 1985 (to 540,000 tonnes), but fell short of the expected targets for those years—535,000 tonnes for 1984 and 545,000 tonnes for 1985.[145] Therefore, from 1981 to 1985, the annual growth rate of the food yield in QN-ĐN was about only 1.4 per cent.

Despite a slight increase in QN-ĐN's staple food production from 1981 to 1985, collective members' incomes from the collective sector deteriorated because many households could not produce more than the quota and the value of their collective workdays was low.[146] For example, in Bình Lãnh collective (Thăng Bình district), staple food production increased slightly, from 2,300 tonnes in 1982 to 2,400 tonnes in 1983 and 2,600 tonnes in 1984, but the value of a workday decreased, from 2 kilograms of paddy in 1982 to 1.3 kilograms in 1983 and about 1.4 kilograms in 1984. Similarly, in Duy An collective no. 1 (Duy Xuyên district), staple food production increased 20 per cent during 1982–84, but the value of a workday deceased from 2.7 kilograms in 1982 to 2.2 kilograms in 1984.[147]

---

143 Vietnamese statistics are often flattering for political and propaganda purposes; however, the numbers can be useful for comparison (Paddy production over the past years, *Quảng Nam-Đà Nẵng*, 14 September 1983, p. 1).

144 Phạm Đức Nam: Kết quả năm 1983 và phương hướng phấn đấu năm 1984 trên mặt trận sản xuất nông nghiệp của tỉnh nhà [Phạm Đức Nam: The results of agricultural production in 1983 and plans for 1984], *Quảng Nam-Đà Nẵng*, 15 October 1983, p. 1.

145 Tổng kết sản xuất nông nghiệp năm 1985, chuẩn bị cho vụ Đông–Xuân tới [Summary of agricultural production in 1985 and preparing for the winter–spring crop], *Quảng Nam-Đà Nẵng*, 21 September 1985, p. 1.

146 The product contract, *Quảng Nam-Đà Nẵng*, 8 November 1984, p. 1.

147 Ibid.

According to *Quảng Nam-Đà Nẵng* newspaper accounts, one reason for the decreased value of a workday in the period 1981–85 was an increase in the state's staple food procurement from the collective sector and unfair terms of trade between agricultural inputs and agricultural outputs, which favoured the former.[148] For example, in QN-ĐN, state food procurement increased from 61,227 tonnes in 1980 to 110,000 tonnes in 1984 and to 120,877 tonnes in 1985, accounting for about 22 per cent of total yield.[149] Another report showed that collective staple food obligations had increased 2.41 times between 1980 and 1984.[150] Meanwhile, the price of paddy was low but the prices of agricultural inputs and other industrial goods were high during 1982–85. Many households therefore could not farm profitably and 'were afraid to invest and expand their production.'[151]

Quảng Nam-Đà Nẵng reports in late 1983 revealed several cases of weak collectives across various districts of the province. For example, an investigation in June 1983 discovered that many collectives in Tam Kỳ district were weak and had veered from party directives. In these collectives, 'blank contracts' were popular at both the collective and the brigade levels. Draught animals had not been fully collectivised, so collectives could not use them. Land and labour were loosely managed, so collective members' earnings from the sector made up only a minor proportion of their total income.[152] A close investigation of Tam Phước collective, one of the weak collectives in Tam Kỳ district, showed that, after adopting product contracts, its collective relations of production weakened. The collective had 1,257 hectares of rice land and 563 hectares of secondary-crop land; however, only half the rice land was used for collective farming. The remainder, especially the secondary-

148 Ibid.
149 Quán triệt nghị quyết hội nghị lần thứ 7 Ban chấp hành trung ương đảng: Ban chấp hành đảng bộ tỉnh quy định phương hướng nhiệm vụ năm 1985 [Full resolution of the 7th Plenum of the Provincial Party Committee: Plans for the year 1985], *Quảng Nam-Đà Nẵng*, 2 February 1985, p. 1; Năm năm phát triển sản xuất nông nghiệp [Five years of agricultural production], *Quảng Nam-Đà Nẵng*, 1 February 1986, p. 2.
150 Summary of three years of implementing the product contract, *Quảng Nam-Đà Nẵng*, 6 July 1985, p. 1; Ban Kinh Tế Tỉnh Ủy QN-ĐN (1985), *Tốc độ khôi phục kinh tế và phát triển xã hội của tỉnh gần 10 năm giải phóng*, [QN-ĐN's Economic Performance over the Past 10 Years], 16 February, Tam Kỳ: Quảng Nam-Đà Nẵng.
151 Summary of three years of implementing the product contract, *Quảng Nam-Đà Nẵng*, 6 July 1985, p. 1. The price ratio of urea fertiliser to paddy in 1983–84 was about 1:2.
152 Củng cố và đưa các hợp tác xã nông nghiệp của huyện Tam Kỳ tiếp tục tiến lên [Solidifying and advancing collectives in Tam Kỳ district], *Quảng Nam-Đà Nẵng*, 4 June 1983, p. 3.

crop land, was used by landowners for their household economy. Therefore, a large proportion of collective members' income came from their own household economic activities.[153]

Collective organisations also suffered losses. According to An Giang's Provincial Committee of Agriculture (*Ban Nông Nghiệp Tỉnh Ủy*), the economic efficiency of collective activities during 1982–84 was so low that many could not even cover their costs. (For example, in 1984, 24 of 40 collectives in Thăng Bình district suffered a loss.) To reduce such losses, collectives in QN-ĐN increased their quotas and the price of agricultural inputs sold to peasants. They also reduced household investment and lowered the value of a collective workday, which is why collectives in the province paid their members low rates for a workday— less than 1 kilogram of paddy.[154]

In November 1984, an investigation into eight collectives in different areas of QN-ĐN found that members in Bình Nguyên's collectives in Thăng Bình district suffered an average loss of 200 kilograms of paddy per hectare. Those in Đại Phước collective in Đại Lộc district lost 400 kilograms per hectare. Members in Tam Nghia collective in Tam Kỳ district lost 123 kilograms per hectare, while Tam Thái collective no. 1 in Tam Kỳ district lost 123 kilograms; Đại Hiệp collective no. 2 in Đại Lộc district lost 148 kilograms; and Bình Lãnh collective in Thăng Bình district lost 210 kilograms. Only in Đại Phước collective no. 1 in Đại Lộc district did collective members show an average profit, of 54 kilograms of paddy per hectare.[155]

After *Đổi Mới* officially began in 1986, the performance of agriculture and collective farming in QN-ĐN dropped alarmingly. The province's staple food production fell from 540,000 tonnes in 1985 to 463,000 tonnes in 1987.[156] For paddy and corn, in particular, production dropped between 1985 and 1988 (see Table 7.1). At a meeting about 'solidifying and strengthening agricultural production relations' in

---

153 Tam Phước củng cố hợp tác xã nông nghiệp [Tam Phuoc solidifies collectives], *Quảng Nam-Đà Nẵng*, 9 July 1983, p. 3. The article did not mention how the other half of rice land had been used. It seems this land was largely under the control of landowners.

154 Ban Nông Nghiệp Tỉnh Ủy QN-ĐN (1984), *Những vấn đề cần giải quyết để phát huy động lực của chế độ khoán mới trong hợp tác xã sản xuất nông nghiệp* [*Some Ideas to Facilitate the Incentives for Product Contracts*], 24 November, Tam Kỳ: Quảng Nam-Đà Nẵng.

155 Ibid.

156 Sơ kết sản xuất nông nghiệp năm 1987, chuẩn bị vụ sản xuất Đông–Xuân tới [Preliminary summing up of 1987 agricultural production and preparing for the winter–spring crop], *Quảng Nam-Đà Nẵng*, 17 September 1987, p. 1.

June 1987—and in the spirit of 'looking the truth straight in the eye' (*nhìn thẳng sự thật*), inspired by the *Đổi Mới* policy—provincial leaders recognised 'some problems and weaknesses' with the product contract system. They admitted that weak collectives were still numerous. Of 270 collectives in the province, 78 were weak (28.9 per cent), 103 were average (38.1 per cent) and 89 were good or advanced (33 per cent). In the midland area, weak collectives accounted for 45.2 per cent of the total.[157] Collectivisation in Thăng Bình district was in an even worse situation: 36 per cent of its collectives were classified as weak, while only 19 per cent were considered good.[158]

Quảng Nam-Đà Nẵng newspaper reports noted several reasons for the decrease in collective farming's performance and peasants' living conditions between 1986 and 1987. First, unfavourable weather affected crop yields. Second was the negative effect of the central government's 'price–wage–currency' reforms in September 1985. In particular, from late 1985, prices across the board in QN-ĐN increased sharply. The price of agricultural inputs increased faster than that of agricultural produce, leading to agricultural produce being sold below cost.[159] Third, the quantity, quality and variety of agricultural inputs were inadequate. In the two price systems (state and free market prices), enterprises serving state farms often sold agricultural inputs on the free market to make a quick profit at the expense of collectives. Meanwhile, collective organisations still lacked economic autonomy.[160] Finally, cadres embezzled, stole collective resources and 'prolonged work and inflated work-points' for non-farming activities.[161]

---

157 Hội nghị củng cố và tăng cường quan hệ sản xuất trong nông nghiệp kết thúc tốt đẹp [The conference on solidifying agricultural production relations produced good results], *Quảng Nam-Đà Nẵng*, 16 June 1987, p. 1.

158 Thăng Bình mở rộng hội nghị củng cố phong trào hợp tác hóa [Thang Binh held a conference on solidifying collectives], *Quảng Nam-Đà Nẵng*, 18 August 1987, p. 3.

159 Tỉnh Ủy Quảng Nam-Đà Nẵng (1987), Nghị quyết của Tỉnh ủy tiếp tục củng cố và tăng cường quan hệ sản xuất, hoàn thiện cơ chế khoán sản phẩm [Provincial Party Committee's resolution on continuing to solidify production relations and perfect the product contract], *Tỉnh Ủy Quảng Nam-Đà Nẵng*, 9 July. According to Nguyễn Khắc Viện, policies relating to the exchange of banknotes and readjustment of prices and wages in 1985 caused the country's hyperinflation. Overall, prices increased 200 per cent in 1985, 550 per cent in 1986 and 400 per cent in 1987. See Nguyễn Khắc Viện (1990), *15 năm ấy: 1975–1990* [*15 Years: 1975–1990*], Hồ Chí Minh: NXB TP, p. 96.

160 Tỉnh Ủy Quảng Nam-Đà Nẵng, Provincial Party Committee's resolution on continuing to solidify production relations; Preliminary summing up of 1987 agricultural production, *Quảng Nam-Đà Nẵng*, 17 September 1987, p. 1.

161 Hoàn thiện cơ chế khoán sản phẩm trong nong nghiệp [Perfecting the product contract system], *Quảng Nam-Đà Nẵng*, 1 March 1988, p. 1.

Table 7.1 Grain production (including paddy and corn) in QN-ĐN, 1976–88

Year	Grain production (tonnes)
1976	154,386
1977	181,687
1978	235,387
1979	282,441
1980	285,426
1981	293,504
1982	330,760
1983	328,166
1984	332,863
1985	358,195
1986	287,362
1987	307,344
1988	299,774

Source: Cục Thống Kê tỉnh Quảng Nam (CTKQN) (2005), *Quảng Nam 30 Năm Xây Dựng và Phát triển* [*Quang Nam's Socioeconomic Development over the Past 30 Years*], Tam Kỳ: Cục Thống Kê tỉnh Quảng Nam, p. 95.

Villagers in Hiền Lộc and Thanh Yên gave several reasons for the low performance of collective farming under the product contract system. One was a decrease in state investment in collective farming, which meant households did not have adequate chemical fertilisers and pesticides. Peasants' inadequate care of fields led to low paddy productivity and, due to a lack of ownership rights, land was overexploited and degraded over time. In addition, the rice seeds were of poor quality—coming from stock people had planted again and again.[162]

Return to household farming

In response to peasants' resistance and the poor performance of collective farming, from late 1986 to 1987, some collectives in QN-ĐN began to experiment with new farming arrangements. For example, when a large number of households returned their contracted land, the managerial board of Bình Tú collective no. 1 in Thăng Bình district decided to implement 'package contracts' (*khoán gọn*) for peasants in the winter–spring of 1986–87. Under this arrangement, the work-

---

162 Author's interviews, 4–19 October 2005, Hiền Lộc.

points system was eliminated and the board announced in advance the cost of inputs, taxes and other fees. After paying these items, peasants were allowed to keep whatever was left. The board faced criticism from higher-level authorities about derailing and destroying socialist production relations; however, the new contracts resulted in peasants who had returned land asking for it back.[163]

Similarly, in Điện Nam collective no. 2 in Điện Bàn district, after falling 117 tonnes below its paddy production quota and 30 per cent of peasant households returning their land, leaders searched for a better farming arrangement in the winter–spring of 1986–87. To encourage peasants to retain their contracted land, the board decided to reward each household by lending it 1.3 *sào* of land for its own use if it also continued farming on its contracted land. Moreover, the collective cadres decided to implement contract no. 100 for only two farming seasons per year and to use 'straight contracts' (*khoán thẳng*) for the third season. Under the straight contracts, peasants knew in advance what they would have to pay the collective. The remainder of their harvest belonged to them, which produced an 'enthusiastic' (*phấn khởi*) response.[164]

Other collectives, such as Hòa Sơn collective in Hòa Vang district and Điện Phước collective no. 1 in Điện Bàn district, also brought in new farming arrangements. A former chairman of Bình Lãnh collective admitted that his collective in the mid-1980s made 'package contracts' with peasants for infertile land they had returned or refused to farm under product contracts.[165] The names of the new arrangements differed from one collective to another, but included 'household contracts' (*khoán hộ*), 'package contracts' (*khoán gọn*) and 'agreement contracts' (*khoán hợp đồng*).[166]

In general, collectives experimenting with new farming arrangements achieved improved results, which came to the attention of provincial authorities. In June 1987, QN-ĐN's leaders held a conference

---

163 Sự thật về cách khoán mới ở Bình Tú 1 [The true story of new contract arrangements in Bình Tu Collective No. 1], *Quảng Nam-Đà Nẵng*, 23 June 1988, p. 4; Chuyện đồng ruộng cuối năm [Collective farming at the end of the year], *Quảng Nam-Đà Nẵng*, 31 December 1987, p. 3.

164 New farming arrangements created new incentives, *Quảng Nam-Đà Nẵng*, 24 December 1987, p. 2.

165 Author's interview, 24 October 2004, Bình Lãnh.

166 Qua một năm cải tiến công tác khoán sản phẩm trong sản xuất nông nghiệp [An evaluation after one year of improving the product contract system], *Quảng Nam-Đà Nẵng*, 30 August 1988, p. 2.

'on solidifying and strengthening agricultural production relations'[167] at which they authorised new farming arrangements by releasing Directive No. 03 (22 June 1987), stressing 'solidifying and strengthening production relations and perfecting the product contract in agriculture'. The directive called for an increase in the economic autonomy of collectives and advocated new farming arrangements in them called the 'agreement contract according to price unit' (*khoán hợp đồng theo đơn giá*). Under these new contracts, collectives had to inform members of their obligations and benefits up front and eliminate widespread subsidies. In addition, under the terms of the new contract, collective members were allowed and even encouraged to buy their own means of production, such as draught animals and small farm machines.[168] The new contracts spread to many collectives in QN-ĐN and, by the winter–spring of 1987, they had been officially adopted by 34 collectives in the province.[169]

Kerkvliet's study on northern Vietnam showed that farming arrangements other than product contracts also prevailed there in many collectives in 1986 and 1987, and some northern provinces, such as Hà Sơn Bình and Vĩnh Phú, approved new farming arrangements in 1987.[170]

An Giang in the Mekong Delta

Like their counterparts in QN-ĐN, many production units in An Giang saw their farming performance improve in the first few seasons of the product contract system. An Giang's staple food production grew from 691,561 tonnes in 1981 to 835,000 tonnes in 1982.[171] However, from 1983 to 1985, when An Giang's authorities pushed the process of 'socialist transformation', the province's staple food production stagnated, and then declined. Table 7.2 shows how paddy production and the amount of cultivated land increased in the early 1980s, but then dropped considerably in 1983 and 1984. Due to a large amount of

---

167 The conference on solidifying agricultural production relations, *Quảng Nam-Đà Nẵng*, 16 June 1987, p. 1.

168 Nghị quyết số 03/NQ-TU: Tiếp tục củng cố và tăng cường quan hệ sản xuất, hoà thiện cơ chế sản phẩm trong nông nghiệp [Resolution No. 03/NQ-TU: Continue to improve and perfect the product contract], *Tỉnh Ủy Quảng Nam-Đà Nẵng*, 22 June 1987.

169 Collective farming at the end of the year, *Quảng Nam-Đà Nẵng*, 31 December 1987, p. 3.

170 Kerkvliet, *The Power of Everyday Politics*, p. 224.

171 Con số niềm tin [The figures and faith], *An Giang*, 20 March 1983, p. 2.

abandoned land and flooding, An Giang's paddy production fell from 820,952 tonnes in 1982 to 792,486 tonnes in 1983 and 725,392 tonnes in 1984.[172]

Table 7.2 Cultivated area of crops and paddy production in An Giang, 1975–88

Year	Cultivated area of annual crops (ha)	Cultivated area of rice paddy (ha)	Paddy productivity (tonnes/ha)	Annual production of paddy (tonnes)
1975	236,594	217,629	2.157	469,426
1976	255,743	220,670	2.249	496,287
1977	278,559	241,593	1.972	476,421
1978	275,980	233,513	1.555	363,113
1979	263,389	231,568	2.271	525,891
1980	329,321	292,374	2.524	737,952
1981	335,092	296,016	2.336	691,493
1982	324,064	283,772	2.893	820,952
1983	325,303	278,652	2.844	792,486
1984	308,153	257,963	2.812	725,392
1985	300,705	263,214	3.451	908,352
1986	312,389	258,805	3.277	848,104
1987	317,139	261,090	3.389	884,834
1988	324,148	262,930	3.729	980,466

Source: Cục Thống Kê An Giang (CTKAG) (2000), *Niên giám thống kê tỉnh An Giang [An Giang Statistical Year Book]*, Long Xuyên: Cục Thống Kế An Giang, pp. 61–75.

According to An Giang's Department of Agriculture, the province's staple food production increased to 923,000 tonnes in 1985; however, this was still short of the target. This increase resulted mainly from an increase in the number of crops per year and the extensive adoption of high-yielding rice. In particular, An Giang expanded the area of land planting two crops per year with high-yielding rice from 34,000 hectares

---

172 Cục Thống Kê An Giang [hereinafter CTKAG] (2000), *Niên giám thống kê tỉnh An Giang [An Giang Statistical Year Book]*, Long Xuyên: Cục Thống Kế An Giang, pp. 61–75. An Giang's staple food yield in 1984 reached about 755,732 tonnes and met only 84 per cent of the target (Ngành nông nghiệp tổng kết công tác năm 1984: Vượt qua khó khăn, toàn tỉnh gieo trồng 300,842 ha [Summing up 1984 agricultural production: Overcoming difficulties to cultivate 300,842 hectares], *An Giang*, 21 February 1985, p. 1).

in 1976 to 180,000 hectares in 1985.[173] Therefore, despite the decrease in An Giang's cultivated area of more than 20,000 hectares from 1982 to 1985, better paddy productivity increased production from 820,952 tonnes in 1982 to 908,352 tonnes in 1985. Therefore, the average growth rate of paddy production in An Giang from 1982 to 1985 was about 3.5 per cent.[174] In assessing agricultural conditions, a 1986 report by An Giang's party executive committee revealed widespread problems:

> In general, agricultural production developed slowly and unevenly. Investment in agriculture did not meet requirements; the price of agricultural produce was still fixed low [gò ép] and was not attractive [to peasants]. Due to agrarian policy shortcomings, some cultivated land was used inefficiently or abandoned. Furthermore, the number of new agricultural machines could not compensate for the damage and loss of old machines.[175]

As in QN-ĐN, despite food production in An Giang increasing slightly during 1982–85 (staple food per person increased from 515 kilograms per year in 1982 to 530 kilograms in 1985), peasants' living conditions did not improve.[176] There were at least three reasons for this. First, the terms of trade between agricultural produce and industrial products (including agricultural inputs) had deteriorated at the expense of the latter. For example, in 1975, 1 kilogram of paddy was worth 1 kilogram of urea or 1.5 litres of fuel. In 1985, 4 kilograms of paddy could buy only 1 kilogram of urea or 1 litre of fuel. Second, the state's food procurement increased considerably between 1982 and 1985. During the period 1983–85 alone, An Giang authorities took 851,000 tonnes of grain (nearly the entire annual output), 30,000 tonnes of beans and sesame, 18,400 tonnes of pork and 21,200 tonnes of fish. Food procurement in 1982–85 increased 28.1 per cent compared with the previous period, 1980–82.[177] Finally, local cadres' embezzlement, theft and poor management (discussed above) and high payments to production units negatively affected peasants' incomes.

---

173 Nguyễn Vũ: Tiếp tục đưa nhịp độ phát triển nông nghiệp lên nhanh hơn [Nguyễn Vũ: Continue to speed up agricultural production], *An Giang*, 24 October 1986, p. 1. Nguyễn Vũ was manager of An Giang's Department of Agriculture.

174 CTKAG, *An Giang Statistical Year Book*, pp. 61–75.

175 Báo cáo chính trị của Ban chấp hành đảng bộ tỉnh An Giang [The political report of Executive Committee of An Giang Provincial Party Committee], *An Giang*, 24 October 1988, p. 3.

176 CTKAG (1986), *Tình hình kinh tế xã hội tỉnh An Giang 1983–1985* [*An Giang's Socioeconomic Situation from 1983–1985*], Long Xuyên: Cục Thống Kê An Giang, p. 7.

177 Ibid., p. 10.

After the completion of collectivisation in 1985, agricultural production in An Giang did not improve. In assessing the province's economic performance in 1986, provincial resolution No. 1/NQ-TU (29 November 1986) revealed:

> The provincial socioeconomic situation was more difficult and complicated than in 1985 due to price–wage–currency adjustments. Some targets were not met; food production fell compared to 1985; more than 10,000 hectares of land were abandoned; farm machines were seriously damaged and lost … economic and social evils, violations of labourers' mastery rights and oppression of the masses become widespread. Especially at the local level, managerial boards of production units committed many serious wrongdoings.[178]

A chairman of An Giang province reflected:

> From 1980 to 1986, due to the consequences of socialist agricultural transformation, forced collectivisation and bureaucratic red tape, food production [in An Giang] stagnated, increasing only slightly, from 741,000 tonnes in 1980 to 855,000 tonnes in 1986. In general, over 10 years after reunification, despite the party organisation and people concentrating on staple food production, it increased only 400,000 tonnes. So, the average annual increase in staple food production was about 40,000 tonnes … In addition, during that time, more than 30,000 hectares of land was abandoned.[179]

A former cadre of An Giang's Committee for Agricultural Transformation listed three reasons for the poor performance of agriculture in the mid-1980s. First, the prices paid for food procurement were low, which discouraged peasants from increasing their production. Second, the state's supply of inputs was inadequate and delivered late, so peasants often 'sowed seeds only' (sạ chay), without using fertilisers, irrigation and other inputs. Finally, the combination of these and other factors meant peasants were not interested in farming (không thiết tha với ruộng đất).[180]

From late 1986 onwards, in the new political atmosphere inspired by the VCP's Đổi Mới policies, moves to strengthen collective organisations in An Giang faced even more challenges. Many peasants took advantage of the spirit of Đổi Mới, which gave the people more freedom to speak,

---

178 Cited in Võ Tòng Xuân and Chu Hữu Quý, *KX Account 08-11*, p. 40.
179 Nguyễn Minh Nhị, *An Giang*, p. 1.
180 Author's interview, 27 June 2005, Long Xuyên.

and sent petitions to ask for return of their former land and machinery and to complain about cadres' embezzlement of collective resources, theft of land and oppression of the masses. Also in the spirit of *Đổi Mới*, An Giang's journalists were given more power to fight 'social evils'. During 1987, journalists exposed many cases of local cadres' misbehaviour, such as embezzling resources, misappropriating peasant land, mismanaging collective funds and oppressing the masses (*ức hiếp quần chúng*). Many production units were also criticised for their poor performance and large debts. By the end of 1987, the total debt of collective organisations in An Giang had reached 10,000 tonnes of paddy.[181]

Assessing collective organisations in September 1987, the chairman of An Giang's agriculture department concluded that weak production units and cadres' malpractices were still widespread. This hindered agricultural production and made peasants feel insecure and discontent. He attributed these problems to hasty collectivisation and a lack of well-trained cadres. He also considered bureaucratic red tape and subsidy mechanisms (*cơ chế quan liêu bao cấp*) harmful to agriculture and especially to collective farming.[182]

The performance of production units continued to deteriorate in the late 1980s along with peasants' living conditions. The *An Giang* newspaper reported in August 1987 that 50 per cent of peasant households had to rely on buying paddy on credit and could pay for it only after the harvest. The article listed three reasons for the fall in peasants' living conditions: first, much of what peasants produced was extracted by state agencies, while agricultural inputs (such as fuel, fertilisers and pesticides) arrived late, were inadequate or were not what was needed. For example, even months after planting paddy, some peasants had not received their agricultural inputs. Second, paddy productivity was severely reduced due to insufficient supplies of agricultural inputs and irrigation. However, peasants were still required to pay for these provisions and to contribute to collective funds. In addition, they had to pay the debts of local cadres and party members. Finally, prices paid

---

181 Peasants need a rational method of agricultural investment, *An Giang*, 23 October 1987, p. 2.
182 Phỏng vấn Nguyễn Vũ: Nhất định khắc phục những yếu kém đưa tập đoàn sản xuất tiến lên một bước [Interview with Nguyễn Vũ: Be certain in correcting shortcomings to advance production units], *An Giang*, 18 September 1987, p. 2. Nguyễn Vũ was the chairman of An Giang's agriculture department.

for paddy were set much lower than those in the free market while the prices of state goods sold to peasants were relatively high. In addition, agricultural taxes were disadvantageous for peasants.[183]

An investigation into peasants' earnings in August 1987 showed that production unit members received an average of 2 *gia* (40 kg) of paddy per hectare. Peasants complained their costs were illogical. They had to pay input costs (the B contract), quotas, transport costs for inputs, support for invalid and martyr families and for irrigation, threshing of paddy, ink and paper and so on.[184] Table 7.3 shows that, because of this cost burden, what remained for each production unit member at the end of a harvest was only 30.2 kilograms of paddy.[185]

Table 7.3 Results and distribution in the average production unit in An Giang in the summer–autumn of 1987

1. Total number of households	115
2. Total number of people	856
3. Total number of workers	459
4. Area of land (hectares)	49.4
5. Output of paddy (kilograms)	123,500
6. Expenditure (kilograms of paddy)	
Land preparation	9,580
Irrigation	2,559
Urea	33,509
Fuel	2,500
Lubricating oil	400
Pesticide	6,420
Paddy seeds	14,820
Fee for pumping water	8,860
NPK fertiliser	8,401
Diesel	4,762
Threshing of paddy	1,880
Managerial fees	617

183 Giá cả thu mua, chính sách thuế nông nghiệp ảnh hưởng đến đời sống của nông dân [Procurement prices and agricultural taxes affect peasants' living standards], *An Giang*, 28 August 1987, p. 3.
184 Mỗi tuần một chuyện: Chuyện ở tập đoàn sản xuất [A story each week: Production unit story], *An Giang*, 28 August 1987, p. 7.
185 Ibid.

Fee for indirect labour	517
Other	2,785
7. Total expenditure (kilograms of paddy)	97,610
8. Remainder for production unit's members (kilograms of paddy)	25,890
9. Paddy income per công (0.1 hectare) of land (kilograms of paddy)	52.4
10. Income per person (kilograms of paddy)	30.2

Source: Mỗi tuần một chuyện: Chuyện ở tập đoàn sản xuất [A story each week: Production unit story], An Giang, 28 August 1987, p. 7.

According to villagers in Long Điền B, during the time of production units, farming achieved poor results and generated low incomes. A formerly landless man in the village recalled:

> In the past [before 1975], a wage earner could get 2–3 kilograms of paddy per day, but farming under the production unit, we got less than 1 kilogram of paddy per day. Before reunification, it was easy to make a living, but after reunification [and until decollectivisation], we worked hard but did not have any surplus; our lives were difficult. The state forced us to accept land but we did not feel happy because farming did not give us good earnings.[186]

Similarly, a former production unit leader in the commune commented:

> After collectivisation, all households here became poor; no-one was able to get rich. Before reunification, people in the Southern Region lived in a market economy so they had comfortable lives. When implementing land redistribution, some households who traded and engaged in non-farming work also accepted land because they feared going to new economic zones … The state saw the failure of collectivisation and changed their policy because they saw that, nine to 10 years after reunification, living conditions of people had been set back [đi thụt lùi].[187]

The return to household farming

After An Giang completed its socialist agricultural transformation, agricultural production faced even more difficulties. Paradoxically, some agrarian policies resulted in outcomes in An Giang that differed from what the VCP and provincial leaders wanted. Land redistribution was aimed at giving land to landless and land-poor households and boosting collectivisation and agricultural production, but, in reality, it benefited

---

186 Author's interview, 3 August 2005, Long Điền B.
187 Author's interview, 5 August 2005, Long Điền B.

local cadres' families and relatives, angered former landowners and caused agricultural stagnation. The non-resident cultivator prohibition and collectivisation had similar consequences. As a result, a large amount of land was abandoned, misused or misappropriated by cadres and government organisations; collectivised farm machines were damaged; and, more importantly, food production deteriorated alarmingly.

In late 1986, An Giang's leaders started to acknowledge and tried to correct these shortcomings of the agrarian policies. To utilise abandoned land and boost agricultural production, provincial leaders decided to cancel the non-resident cultivator prohibition and to grant more land to households who had greater farming capacity.[188] On 19 February 1987, provincial leaders issued Decision No. 93-NQUB, aimed at correcting the shortcomings of previous land redistributions. To protect agricultural machines from further damage, they discontinued the collectivisation of peasants' machines and urged collective organisations to return machines to their previous owners (peasants).[189] These policies were mainly aimed at improving agricultural performance, but they triggered peasants' moves to reclaim their land and machines.

Despite land conflicts disrupting agricultural production in rural areas, in 1987, production recovered thanks to the corrective measures.[190] Inspired by these positive effects and finding that most production units in the province were, in fact, problematic, in early 1988, An Giang's leaders started to question the direction of collective organisations. In January 1988, Võ Quang Liêm, the vice-secretary of An Giang's party committee, admitted:

> Collective organisations are now unsuitable because they are inefficient in terms of production and their managerial bodies are bulky and unnecessary. Collective organisations manage poorly and commit numerous wrongdoings, which hinder agricultural production and negatively affect the living conditions of peasants.[191]

---

188 Officials' meeting thoroughly resolves 1987 resolution, *An Giang*, 22 December 1986, p. 1.

189 In collectivising peasants' farm machines, production units paid the owners in instalments the remaining value of the machines; however, production units often delayed or evaded these payments (Cùng cố và cải tạo máy nông nghiệp, xay xát [Improving and renovating the management of agricultural machines], *An Giang*, 22 December 1986, p. 2).

190 Võ Tòng Xuân and Chu Hữu Quý, *KX Account 08-11*, p. 47.

191 Võ Quang Liêm: Vấn đề cùng cố, nâng chất các tập đoàn sản xuất [Võ Quang Liêm: The matter of solidifying and upgrading production units], *An Giang*, 15 January 1988, p. 1. Võ Quang Liêm was the vice-secretary of An Giang's party committee.

He also argued that, given current production conditions in which farming required a lot of manual work, it was necessary to consider households as basic units. Authorities should grant long-term land use for households and reduce staff on managerial boards to only one or two cadres. Peasants should be allowed to select freely the best farming services available.[192] The provincial resolution of March 1988 called for a redefinition of the objectives of agricultural transformation. It argued that the main objective of the transformation was to facilitate production; however, in the past, An Giang's authorities had misunderstood this objective and 'coerced peasants into joining collective organisations even though it was supposed to be voluntary'. As a result, 'production stagnated; living conditions of peasants were difficult ... [and a] new class of oppressors and exploiters had appeared'—local cadres, mainly of collective organisations.[193]

It is worth noting that the debate about the shift in agrarian policy in An Giang took place before the VCP released Resolution No. 10 (5 April 1988), which officially endorsed the reallocation of land to peasant households to use for 15 years and fixing the quota for five years.[194]

The shift in national policy and the return to household farming

Under the product contract system, especially in the later stages, the deterioration in agricultural production and the performance of collective organisations occurred in almost all provinces of Vietnam. According to a VCP report, product contracts only slightly boosted agricultural production in the period 1981–85, while after 1986, contracts lost their positive effect and food production stagnated (see Table 7.4).[195]

---

192 Ibid.
193 Xác định lại mục đích cải tạo nông nghiệp [Redefining the objectives of agricultural transformation], *An Giang*, 4 March 1988, p. 1.
194 Ban Tuyên Huấn Trung Ương, *The Party's Response to Urgent Land Problems*, pp. 99–100.
195 Bộ Nông Nghiệp (1990), *Dự thảo tổng kết 3 năm thực hiện nghị quyết 10 10 của Bộ chính trị về đổi mới quản lý kinh tế nông nghiệp* [*A Draft Summing Up of the Three-Year Implementation of Resolution No. 10*], 10 December, Hà Nội: Bộ Nông Nghiệp, p. 1.

Table 7.4 Vietnam's staple food production, 1981–87

Year	1981	1982	1983	1984	1985	1986	1987
Staple food production (millions of tonnes, paddy equivalent)	15.0	16.8	16.9	17.8	18.2	18.3	17.5

Source: Bộ Nông Nghiệp (1990), *Dự thảo tổng kết 3 năm thực hiện nghị quyết 10 của Bộ chính trị về đổi mới quản lý kinh tế nông nghiệp* [*A Draft Summing Up of the Three-Year Implementation of Resolution No. 10*], 10 December, Hà Nội: Bộ Nông Nghiệp, p. 2.

According to researcher Nguyễn Sinh Cúc, a decrease of 0.8 million tonnes of food in 1987 compared with 1986, accompanied by a population increase of 1.5 million, caused a sharp decrease in staple food per capita, from 300.8 kilograms per year in 1986 to 280 kilograms per year in 1987—the lowest figure since 1981. In collective organisations, peasants' income accounted for about 20 per cent of the quota. In 21 Vietnamese provinces (from Bình Trị Thiên province northward), 39.7 per cent of rural people suffered severe hunger between harvests (*nạn đói giáp hạt*).[196]

Faced with falling living conditions like their counterparts in QN-ĐN and An Giang, peasants elsewhere in Vietnam were fed up with collective farming. Even in 'good' collectives, peasants began to return contracted land. As a result, land in widespread locations was abandoned and peasants' debts increased over time. In response to the situation, some collectives tried to experiment with 'package contracts', which some local authorities authorised.[197] Despite criticism by party officials and analysts, 'package contracts' or 'household contracts' gradually gained the approval of authorities. According to Ben Kerkvliet, by September 1987, farming arrangements other than product contracts prevailed in more than 70 per cent of the collectives in Vietnam. Finally, in April 1988, the party's political bureau released Resolution No. 10, stressing 'the renovation of agricultural economic management', which implicitly endorsed previous practices and marked the beginning of decollectivisation in Vietnam.[198]

Resolution No. 10 was aimed at unleashing the production capacity of agriculture and shifting it to commodity production, by giving collective organisations and peasant households more autonomy in

---

196 Nguyễn Sinh Cúc, *Agricultural and Rural Development in Vietnam*, p. 47.
197 Bộ Nông Nghiệp, *A Draft Summing Up*, p. 1.
198 Kerkvliet, *The Power of Everyday Politics*, pp. 224, 227.

production. To encourage peasant households to increase production, land was allocated to them for longer-term use (15 years) and quotas were fixed for five years.[199]

VCP leaders did not intend to dismantle collective organisations; however, in the context of a market-oriented economy, after implementation of Resolution No. 10, peasants gradually became independent of collective organisations and they gradually lost their purpose and were dismantled or were changed to farming-service organisations in the early 1990s. The peasant household finally became the basic production unit in rural areas of Vietnam.

Conclusion

In an effort to save collective organisations and improve their performance, in 1981, the VCP released Directive No. 100. The hope was to reduce peasants' and local cadres' problems and to strengthen collective organisations. Even though product contracts immediately improved the performance of collectives and boosted agricultural production, they did not solve the long-term struggle between peasants and local cadres over land, labour and other resources.

Although there were many campaigns to correct cadres' problems, both in QN-ĐN and in An Giang, performance did not improve. Local cadres often took advantage of their position to steal state, collective and peasant resources. Land redistribution, the non-resident cultivator prohibition and collectivisation in An Giang were all aimed at eliminating the old exploitative class, but, in reality, the policies created a new exploitative class—namely, local cadres.

During the product contract system, collective farming essentially replaced commercial farming and the diverse rural economy of An Giang. This is why villagers in An Giang displayed behaviour comparable with their counterparts in QN-ĐN. Villagers in both places tried their best to minimise the disadvantages of collective farming to enhance their own survival and livelihoods. For example, while villagers in QN-ĐN tried their best to enlarge their household economies by capturing collective resources, land and labour, villagers

---

199 Ban Tuyên Huấn Trung Ương, *The Party's Response to Urgent Land Problems*, pp. 81–123.

in An Giang tried their best to ensure their livelihoods by doing wage work and using collective resources for their daily needs. Both tried to avoid paying debts and fulfilling their obligations to the collective; they returned land or abandoned it when they saw that collective farming was unprofitable. All of these behaviours had a huge adverse effect on the survival of collective organisations.

The combined effect of the peasants' and local cadres' practices significantly contributed to the poor performance of collective farming and the failure of the product contract system, which the VCP had expected to improve collective farming. The output of staple food decreased alarmingly after mid-1985 and, in response, local cadres and authorities in QN-ĐN, An Giang and elsewhere experimented with new farming arrangements. When Vietnam faced a food crisis in the late 1980s, the VCP finally gave up on the official system and endorsed new local arrangements, which marked the beginning of decollectivisation and a return to household farming in Vietnam.

8

Conclusion

This book has investigated why socialist collectivisation failed in south Vietnam after 1975. After the country was reunified, the Vietnamese state attempted to implement collectivisation policies similar to those that had been applied in the north of the country during the Vietnam War. Despite the strong will of the new regime to implement this central pillar of socialist ideology, collectivisation was not realised uniformly; it was misapplied and subverted; and, after only 10 years, it was annulled as policy. This set of failings is somewhat of a puzzle for it occurred well before the collapse of the Soviet Union, when socialist ideology in Vietnam was at its peak and was manifest in many aspects of social policy. It also took place when the state was at its most confident and resolute after the successful reunification of the country. Focusing on two case studies (Quảng Nam-Đà Nẵng province in the Central Coast region and An Giang province in the Mekong Delta) and based on an extensive review of the evidence, this study suggests that the reasons for variations in policy implementation, and the failure and reversal of policy, were twofold: regional differences and local politics.

The book shows that, at the time of the reunification in 1975, there were significant differences between the Central Coast and the Mekong Delta in terms of the consequences of the country's wars, the impacts of previous agrarian reforms, the natural and socioeconomic conditions and the social structure of rural communities. In Quảng Nam-Đà Nẵng (QN-ĐN) in the Central Coast, prolonged war had disrupted or destroyed any positive effects of previous land reforms and development carried out by either Saigon's government or the National

Liberation Front (NLF). It was the war rather than previous land reforms that had transformed the local land tenure system, by causing landlords and a large proportion of the rural people to abandon their houses and land and live in enclosed camps. After the war, many landlords did not return, while others returned but could not reclaim or restore all of their land. Large areas were overgrown with weeds and seemed to have come under a kind of communal ownership. People restored any plot they liked as if it were their own. Thus, the war had changed the land tenure system and flattened the structure of rural communities, leaving the society relatively homogeneous. In addition, after the war, the peasant economy was mainly subsistence-focused. Most peasants were rendered poor and were mainly concerned with producing enough food for their families. As in the north, peasant communities in QN-ĐN continued to practise labour exchange and possessed a strong collective sense in accordance with traditional thinking about reciprocity and mutual assistance, especially during difficult times.

Meanwhile, soon after reunification, the agricultural sector in An Giang in the Mekong Delta had reached a higher level of economic development than that in the Central Coast and in the north in the 1950s. The social structure and rural economy in the Mekong Delta were more diverse. Previous agrarian reforms carried out by the Việt Minh, various South Vietnamese governments or the NLF had significantly changed the land tenure system and boosted commercial agriculture. By 1975 in the Mekong Delta, 70 per cent of the rural population were middle peasants who owned 80 per cent of the cultivated land, 60 per cent of the total farm equipment and 90 per cent of draught animals. Market relations and individual land tenure had been well established. The landless and land-poor could make a decent living by working as agricultural labourers, engaging in small trading or pursuing other off-farm economic opportunities. Many middle peasant farmers owned their own machinery, cultivated land beyond their residential area and, unlike those in the Central Coast, engaged in commercial rather than subsistence farming. Therefore, concepts of mutual aid, labour exchange and reciprocity were unpopular and occurred only among members of an extended family. In general, unlike the agrarian sector in the Central Coast and in northern Vietnam in the 1950s, the agrarian sector in the Mekong Delta was dominated by middle peasants who engaged largely in commercial agriculture and who wanted to continue to farm their own land and sell their own crops.

Differences in revolutionary influence also contributed to the disparities between the two regions. Large parts of the rural areas of QN-ĐN and the wider Central Coast region were under the influence of the Việt Minh during the war with France (1945–54) and then under the NLF during the war with America (1954–75). Therefore, peasants in these regions had a stronger relationship with the Việt Minh and the NLF and were more familiar with their respective political and economic policies. During the American war, the NLF was able to recruit a large number of revolutionaries who operated locally or were sent to the north for training. After the war, thanks to the considerable number of local revolutionaries who survived or returned from northern Vietnam, the Central Coast did not face a huge problem filling local government and party positions. These cadres were familiar with, enthusiastic about and committed to the post-1975 socialist transformation policies, at least in the first few years. Meanwhile, in An Giang during the American war, the area and therefore the population under the influence of the revolutionaries was reduced, local networks of revolutionaries were destroyed and many revolutionaries were killed or deserted. In many locations, no Communist Party cells operated until after reunification and, even then, they were few in number and relatively weak. The new local-level authorities had to recruit new cadres, a majority of whom were not ex-revolutionaries and were therefore not familiar with socialism and were unenthusiastic about the socialist transformation of agriculture and collectivisation.

Despite these regional disparities and the high level of commercial agriculture in the Mekong Delta, Vietnamese Communist Party (VCP) leaders decided to impose on the south the northern model of socialist agricultural transformation, which was considered a central pillar of socialist construction. As in the north and in other socialist countries, this agrarian reform consisted of two key components: land reform and collectivisation, with the former an essential step to prepare for the latter. Regardless of the existing shortcomings and disappointments of collective farming in the north, the VCP believed that collectivising agriculture in the south was the only way to modernise it, eliminate exploitation, support industrialisation and improve peasants' living standards. VCP leaders were apparently propelled by a commitment to building socialism and socialist large-scale production and a strong belief in their own capacities.

The VCP strongly believed the socialist transformation of agriculture in southern Vietnam could be completed by 1980. The socialist project, however, was not driven solely by the strong will and high ideology of top-level leaders; it also greatly depended on local-level conditions and politics, the main actors of which were ordinary peasants and local cadres. As it happened, the results of the post-1975 agrarian reforms varied from region to region. The implementation of postwar economic restoration measures, land reform and collectivisation was rapid in QN-ĐN and other provinces in the Central Coast. From 1975 to 1978, authorities in QN-ĐN were able to accomplish most of the preparatory measures for collectivisation, such as land redistribution, irrigation, field transformation and the establishment of simple collective organisations and pilot collectives. Authorities in QN-ĐN also met the central government's target to collectivise farming by 1980.

In contrast, authorities in An Giang and elsewhere in the Mekong Delta encountered major difficulties, and to the extent that socialist transformation occurred, it took many years and much effort to complete. Many policies—such as land redistribution and the building of interim collective organisations and pilot collectives—failed to reach the central government's targets and expectations due to strong peasant resistance and the inadequate commitment of local cadres. Local authorities failed to fully establish northern-style collective farms and had to modify and reduce the scale and socialist characteristics of their collectives to ease local resistance. Despite these compromises, collectivisation in the region accounted for less than 10 per cent of agricultural land and peasant households in 1980.

In the period 1981–85, having failed to achieve their targets, the VCP continued to press local authorities in the Mekong Delta to complete socialist transformation. Central Directive No. 100 (January 1981)— which called for the replacement of the work-points system with the product contract system, in which each household farmed separately on their contracted land—collectivisation in the region faced weaker peasant resistance and moved faster than before. Under pressure from the central government, authorities in An Giang and elsewhere in the delta rushed to carry out agricultural transformation, especially from 1984 to 1985. Many collective organisations were hastily established by 'just signing names', and land redistribution divided a commune's land equally between members. By February 1985, the province was declared halfway towards completing collectivisation. By April 1985,

An Giang announced completion of basic collectivisation, accounting for 80 per cent of the province's agricultural land—a minimum index for success. Thus, although the VCP announced the completion of socialist agricultural transformation in the Mekong Delta in the mid-1980s, many collective organisations essentially existed only on paper and fell short of expectations.

Despite encountering great difficulties, VCP leaders were resolute and persisted with the attempt to carry out socialist agricultural transformation and build socialism in the rural south. Only by modifying their policies to ease local resistance were VCP leaders able to establish collective organisations, but still they failed to realise their policy objectives. Collectives became sites of constant struggle between peasants, local cadres and state agencies over land, production and distribution. While VCP leaders were able to force villagers into collective structures, they could not direct peasants and local cadres to behave according to their expectations. This is why collective farming performed very poorly in the Mekong Delta and in the Central Coast region.

During the work-points system (1978–81), villagers in the Central Coast merely went through the motions, trying their best to optimise their work-points rather than the quality of production. In other words, they ended up doing collective work carelessly and deceitfully. Many tried to plunder collective resources and invested most of their energy in their own household economy. Few took care of collective property or worked as enthusiastically as authorities wanted. Meanwhile, the better-off villagers in An Giang tried their best to evade collective farming altogether. To avoid any political disadvantage, some joined production units but did not seriously undertake collective work. Some 'kept one foot within and the other foot outside' the production unit to make a living. Some sent their children or auxiliary labourers to do collective work while they, as their household's main labourers, worked for themselves. Many did collective work carelessly and sluggishly and did not care much about collective property. Although the behaviour of peasants in QN-ĐN and An Giang was quite different, in both places, the main objectives were to minimise the disadvantages of the system and maximise the benefits to themselves. The aggregate of these individual actions contributed significantly to the poor performance of collective farming.

Other key factors contributing to the poor performance of collective farming were local cadres. Despite being loyal to the VCP's agrarian policies, several local-level cadres in QN-ĐN took advantage of their position for personal gain, at the expense of the collective and the overall purpose of the reform. They strictly controlled peasants' economic activities while managing collectives poorly; they embezzled a considerable amount of agricultural inputs and produce. Some assigned tasks and gave work-points to members at their own discretion or prolonged tasks and inflated work-points to favour their fellow villagers or relatives at the expense of others. Some were prejudiced, bureaucratic, autocratic and patriarchal towards members; their behaviour contradicted the authorities' dictum that 'the collective is home and its members are the masters'.

Meanwhile, local cadres in An Giang were unenthusiastic about agricultural transformation and collective farming policies. Faced with peasant resistance, they were reluctant to carry out policies forcefully according to the official blueprint and instead modified policies to accommodate local concerns. Some used specific parts of the national policies, such as the 'positive and firm principle' of collectivisation, to delay the process or let it drift. Compared with their counterparts in the Central Coast, cadres in An Giang allowed peasants more freedom in selecting whether or not they joined collective farming or participated in collective work; however, they managed collective property poorly. *An Giang* newspaper accounts revealed numerous cases of local cadres' sloppy management of production, theft of collective inputs, cash and peasants' work-points, misappropriation of peasant land and property and bullying of the masses. Cadres colluded with merchants, most of whom were their relatives, which contributed to inflation and aided the survival and expansion of the black market. These were the very things the VCP leaders were trying to control and eliminate.

As in north Vietnam, everyday peasant politics and local cadres' malpractices in both QN-ĐN and An Giang provinces during 1979–81 significantly affected the performance of collective farming. Collective farming in both provinces, as elsewhere in Vietnam, performed poorly and food production deteriorated, falling short of official expectations. Despite authorities in both regions putting great effort into correcting peasants' and cadres' negative practices, these behaviours increased over time. In response to the deteriorating performance of collective farming and the steady fall in the country's food production, the VCP decided to abandon the work-points system and introduced a new farming arrangement, the product contract system, which was intended to reduce poor practices and motivate villagers to work enthusiastically and responsibly. The product contract system helped improve the performance of collective farming in both provinces for a few years only, and failed to solve the long-term struggles between peasants and local cadres about land, labour and other resources.

During the product contract period (1981–88), villagers in QN-ĐN tried their best to enlarge their household economies by encroaching on collective resources such as land, labour and agricultural inputs at the expense of the collective economy. Despite authorities expecting them to put collective and state interests first, villagers prioritised their own interests. When they failed to produce more than their quota or faced subsistence shortages, many refused to pay their debts to collectives or fulfil their obligations to the state. In the later stages of the product contract system, when it became clear collective farming was less profitable than outside opportunities, many peasants in QN-ĐN decided to accept less contracted land or even abandoned collective land to make a living elsewhere. This had a huge impact on the performance of collective farming in that province.

By contrast, in An Giang, collectivisation during the product contract period transformed the Mekong Delta's commercial agriculture into subsistence farming. Some landowners who lost their land during land redistribution were disappointed and gave up farming or did just enough to subsist. Many land recipients farmed poorly because they did not know how to farm, lacked incentives or had inadequate capital and were not provided enough help by production units; many put a considerable amount of time and effort into working for wages to supplement their livelihoods. Some sold state agricultural inputs to meet their daily needs rather than investing them in the contracted fields. Like their counterparts in QN-ĐN, in the later stages of the product contract system, many An Giang villagers were in debt to production units, and many refused to repay these debts. Some decided to abandon, transfer or even sell their redistributed land to others.

VCP leaders believed product contracts would reduce the number of problems associated with cadres' malpractice by increasing their responsibility for managing certain phases of collective farming. However, despite numerous campaigns during 1981–88 by the authorities in QN-ĐN and An Giang, aimed at improving the quality of local cadres and correcting and cracking down on their poor behaviour, these problems did not disappear, and in fact increased over time. Local cadres in QN-ĐN tended to shift their responsibility on to villagers by using 'blank contracts' that required villagers to do most phases of farming. Cadres often failed to fulfil their duties, such as spraying pesticides or watering fields on time. They also embezzled scarce collective resources over which they had control, such as agricultural inputs and collective property. From the mid-1980s, when Vietnam adopted a multisectoral market economy, cadres tended to relax their management of collectives and take advantage of opportunities in the free market. For instance, they sold scarce fertilisers on the free market for personal gain—to such a degree that many collectives in Thăng Bình and Quảng Nam did not have enough for their own members in 1987.

Meanwhile, in An Giang, local cadres were guilty of numerous malpractices from 1981 to the late 1980s. Many exploited their positions to steal collective agricultural inputs and funds, and most owed large debts to the state and collectives. Many cadres owed tonnes of paddy and accounted for a majority of the total debt in the province. Although authorities in both places put great effort into correcting and punishing such activities, this sort of behaviour became more prevalent. In addition, local cadres misappropriated a considerable amount of peasants' land, which was supposed to be redistributed to landless and land-poor households. Most production units in An Giang did not operate according to the official product contract system. Rather, they divided land among households to be farmed individually but retained control over household production, distribution and marketing. Production unit cadres monopolised farming services and served members poorly while overcharging them for the cost of these services.

In general, the widespread malpractices of peasants and local officials were at odds with VCP leaders' requirements and contributed to the poor performance of collective farming and the eventual derailment of many national agrarian policies. Collective farming failed to increase productivity or improve peasants' living standards. Collectivisation also aimed to eliminate exploitation but, in reality, it merely created a new class of exploiters in rural areas: collective and production unit cadres. Land redistribution was supposed to benefit the landless and land-poor, but failed to do so. Rather, it largely benefited local cadres and their relatives. The non-resident cultivator prohibition enabled collectivisation but, in turn, significantly hindered peasants' production capacity and commercial agriculture in An Giang.

The failure of collective farming in Vietnam was manifest well before the collapse of the Soviet Union and before Vietnam's withdrawal from Cambodia. Staple food production in QN-ĐN, An Giang and elsewhere in Vietnam declined alarmingly between 1985 and 1987. The living conditions of villagers also deteriorated over time. Fed up with collective farming, many villagers decided to abandon or return land to collectives, especially when Vietnam adopted a market economy under the Đổi Mới policy in 1986. To encourage peasants to farm their collective fields, some local authorities in QN-ĐN tried new farming arrangements as an alternative to the product contract system. Authorities in An Giang also recognised that most production units were inadequate in quality and that collective farming had failed to improve peasants' living conditions. To increase food production, they tried to correct the shortcomings of socialist transformation by allowing peasants to farm outside their villages and returning some land to productive landowners. These practices happened before national leaders launched a major change to their agrarian policy in 1988.

The poor performance of collective farming and the deteriorating living conditions were not confined to QN-ĐN and An Giang, but occurred in most parts of Vietnam during 1985–87. Villagers were hungry in many locations; they accepted less contracted land and some even abandoned land; and their debts increased over time. In response, many locations tried new farming arrangements to deal with their problems. By September 1987, more than 70 per cent of collectives in Vietnam were using farming arrangements other than the product contract. Realising they would not be able to reverse the situation, the VCP in April 1988 released Resolution No. 10, which endorsed the new local arrangements. The resolution marked a new era in Vietnam's agricultural development: the return to household farming.

In summary, central to the failure and then modification of national agrarian policies in southern Vietnam post 1975 were the widespread practices of peasants and local officials that were often at odds the VCP's expectations of 'the new socialist'. Peasants tried their best to pursue their own household economic interests rather than collective ones. Local cadres often took advantage of their position to benefit themselves rather than to serve the people, the collectives and the state. Despite numerous official campaigns to correct and crack down on such bad behaviour and even change national policies to accommodate local concerns, these problems did not disappear, but in fact increased. The ultimate consequences were the inefficiency of collective farming, severe food shortages and an economic crisis that eventually forced the VCP to accept and endorse the farming arrangements that villagers and cadres had initiated to deal with their own local problems.

My findings on collective farming in QN-ĐN and An Giang in southern Vietnam reinforce Ben Kerkvliet's proposition about the power of everyday politics.[1] Despite differences in the form and degree of peasant action, their everyday politics had a huge impact on the performance of collective farming and contributed to the failure and change of national agrarian policies. Moreover, it is clear that socialist agrarian reform faced stronger resistance from peasants in the Mekong Delta than in the Central Coast and the north of the country. Resistance came not only from many landowners, but also through a lack of collaboration from landless households. Peasant resistance in the Mekong Delta took various forms, from subtle everyday politics to open and confrontational resistance. However, like ordinary people, many land-rich and upper–middle peasants resisted state policies individually, rather than mobilising others around them to exercise social control together. Only in favourable conditions were these peasants able to use kinship, informal social networks, local institutions and various other measures to evade or make use of state policies for their own gain. For example, after An Giang authorities' Decision No. 303/QD-UB to correct the shortcomings of land policies, from 1988 to the early 1990s, many upper–middle and middle peasants were able to take advantage of this favourable policy to retrieve their previous landholdings.

One might have expected religious factors to be significant in understanding the course of collectivisation in southern Vietnam. Followers of the Hòa Hảo religion, who were prominent in many parts of An Giang province, might have been obstacles for the VCP's agrarian policies. However, I found that the policies encountered major resistance in the Southern Region regardless of people's religious affiliation. The VCP's plans faced even more problems in NLF-influenced areas

---

1   Kerkvliet, *The Power of Everyday Politics.*

such as Tân Hội commune (Cai Lậy, Tiền Giang province) and some parts of Đồng Tháp province. In Long Điền B, Chợ Mới in An Giang, land redistribution and collective farming encountered the same levels of resistance in areas where Hòa Hảo predominated as in areas where the majority of people were Catholic. When interviewing villagers there, I found that Hòa Hảo and non–Hòa Hảo followers had similar views and experiences of the post-1975 agrarian reforms and similar justifications for their behaviour. I found barely any villagers who used their religion to justify their resistance to collective farming. Chợ Mới district was a Hòa Hảo stronghold, but also the first district in An Giang to complete collectivisation. In general, collective farming encountered problems regardless of whether or not the population was Hòa Hảo. In QN-ĐN, villagers in many locations were not particularly religious, but collective farming ran into trouble there, too. It is therefore likely that religion is not an important factor in understanding the course of collectivisation in An Giang and QN-ĐN.

One might wonder whether struggles between villagers and state agencies over land and other agrarian issues have abated since the reestablishment of household farming. It seems such struggles are not over. Land redistribution, which VCP leaders initially considered a temporary measure towards collectivisation, turned out to be a source of long-term tension and struggle between the party and southern society. Land reform and struggles over it occurred from 1975 to the late 1980s, continued to be a hot issue in the early 1990s and remain so today. For example, despite authorities in An Giang dealing with more than 30,000 peasant complaints in 1988–90, a large number of land conflicts have still not been resolved. Unable to settle persistent and widespread land disputes, An Giang authorities decided in the early 1990s to stop dealing with such matters—a decision that angered many villagers who had not yet regained their lost land. Meanwhile, new land conflicts have emerged since the reestablishment of household farming, especially since the late 1990s, as Vietnam's urbanisation and industrialisation have intensified. State agencies have often taken over villagers' fields without proper compensation. Local cadres across all regions of Vietnam continue to abuse their power to misappropriate villagers' land for their personal benefit. These phenomena have exacerbated rural land conflicts.

In recent years, hundreds of villagers from different regions of Vietnam, disillusioned with local government, have gathered in Hà Nội and Hồ Chí Minh City to demand the central government resolve their land disputes.[2] Some of these disputes have their origins in the post-1975 land redistribution, while others have resulted from the recent process of urbanisation. Villagers' demonstrations have become a hot issue in Vietnam today. In other words, land will likely continue to be a source of rural conflict and political discontent in Vietnam in the coming years.

---

2   BBC Vietnamese (2016), Dân nhiều vùng lên Hà Nội biểu tình đòi đất [Many people in Hanoi protest about land], *BBC Vietnamese*, 21 January, available from: www.bbc.com/vietnamese/forum/2016/01/160120_quynhchau_land_protests (accessed 4 October 2017).

Bibliography

An Giang. (1980). An Giang vững vàng đi tới [An Giang is doing well], *An Giang*, 6 January, p. 1.

An Giang. (1980). Đại hội của trí tuệ tập thể và niềm tin thắng lợi [A meeting with collective wisdom and faith in victory], *An Giang*, 6 January, p. 1.

An Giang. (1980). Đẩy mạnh củng cố và tiếp tục phát triển tập đoàn [Intensifying solidification of production units and extending more], *An Giang*, 20 January, p. 3.

An Giang. (1980). Vụ sản xuất đầu tiên của tập đoàn sản xuất Phú Thượng [The first crop of the Phú Thượng production unit], *An Giang*, 13 March, p. 2.

An Giang. (1980). An Giang đẩy mạnh công tác chống tiêu cực [An Giang speeds up the fight against negativism], *An Giang*, 8 June, p. 2.

An Giang. (1980). Tăng cường chỉ đạo công tác chống tiêu cực [Intensifying the fight against negativism], *An Giang*, 8 June, p. 2.

An Giang. (1980). Vì sao giá lúa leo thang [Why rice prices escalate], *An Giang*, 27 October, p. 2.

An Giang. (1980). Chuyện to nhỏ: Ăn xén của dân [Pilfering people's resources], *An Giang*, 23 November, p. 3.

An Giang. (1980). Phú Tân đẩy mạnh phong trào hợp tác hóa nông nghiệp [Phú Tân intensifies collectivisation], *An Giang*, 27 November, p. 1.

An Giang. (1980). Cải tạo nông nghiệp ở Thoại Sơn [Agricultural transformation in Thoại Sơn], *An Giang*, 7 December, p. 2.

An Giang. (1980). Về thăm tập đoàn số 2 Mỹ Lương [A visit to Production Unit No. 2 in Mỹ Lương], *An Giang*, 7 December, p. 2.

An Giang. (1981). Trong tháng 12, 1980 tỉnh phát triển thêm được 14 tập đoàn sản xuất [In December 1980, the province established 14 more production units], *An Giang*, 11 January, p. 2.

An Giang. (1981). Đạo tạo cán bộ cốt cán cho các tập đoàn sản xuất và hợp tác xã [Training key cadres for production units and collectives], *An Giang*, 1 February, p. 1.

An Giang. (1981). Chuyển biến mới ở HTX Tây Huê [Good progress in Tây Huê collective], *An Giang*, 7 June, p. 2.

An Giang. (1981). Đẩy mạnh công tác cải tạo nông nghiệp [Speeding up agricultural transformation], *An Giang*, 7 June, p. 1.

An Giang. (1981). Phong trào hợp tác hóa tiếp tục đi vào chiều hướng ổn định và phát triển theo hướng phương châm tích cực và vững chắc [Collectivisation continues to progress positively and firmly], *An Giang*, 7 June, p. 2.

An Giang. (1981). Kiếu nại ruộng đất [Petition on land], *An Giang*, 8 June, p. 4.

An Giang. (1981). Tập đoàn sản xuất I, khóm Châu Long 4 vững bước tiến lên [Production Unit No. 1, Châu Long 4 Subcommune is progressing], *An Giang*, 9 August, p. 2.

An Giang. (1981). Vài nét về những kho chứa lúa ở Thoại Sơn [Some problems with rice stores in Thoại Sơn], *An Giang*, 23 August, p. 3.

An Giang. (1981). Vài nét về một tập đoàn yếu kém [Some portraits of a weak production unit], *An Giang*, 6 September, p. 2.

An Giang. (1981). Xã luận: Công tác cải tạo nông nghiệp tỉnh An Giang [The editorial: Agricultural transformation in An Giang], *An Giang*, 6 September, p. 1.

An Giang. (1981). Phong trào hợp tác nông nghiệp ở An Giang từng bước được củng cố đi lên [Collectivisation in An Giang has progressed], *An Giang*, 18 November, p. 1.

An Giang. (1981). Xã luận: Ra sức phấn đấu đưa phong trào cải tạo xã hội chủ nghĩa đối với nông nghiệp ở tỉnh ta tiến lên một bước mới [The editorial: Do the best to take collectivisation in An Giang one step forwards], *An Giang*, 18 November, p. 1.

An Giang. (1982). Phấn khởi với cách khoán mới [Enthusiasm with the product contract], *An Giang*, 14 March, p. 2.

An Giang. (1982). Vụ lúa khoán đầu tiên ở thị xã Long Xuyên [The first contracted rice crop in Long Xuyên town], *An Giang*, 14 March, p. 1.

An Giang. (1982). Trả lời bạn đọc về việc điều chỉnh ruộng đất [Answering readers' questions about land redistribution], *An Giang*, 4 April, p. 1.

An Giang. (1982). Vụ lúa khoán ở tập đoàn 3 Tây Khánh B [The results of contracted rice crops in Production Unit No. 3 in Tây Khánh B Commune], *An Giang*, 18 April, p. 3.

An Giang. (1982). Kết quả khoán ở Long Điền B [The results of the product contract in Long Điền B], *An Giang*, 2 May, p. 3.

An Giang. (1982). Huyện Chợ Mới áp dụng khoán sản phẩm có kết quả [The product contract in Chợ Mới brings about good results], *An Giang*, 9 May, p. 3.

An Giang. (1982). Kết quả tốt đẹp của khoán sản phẩm trong nông nghiệp [The product contract brings about good results], *An Giang*, 23 May, p. 1.

An Giang. (1982). Khoán sản phẩm cuối cùng đến người lao động, một hình thức thích hợp mang lại nhiều kết quả to lớn [The product contract is suitable and brings about good results], *An Giang*, 30 May, p. 1.

An Giang. (1982). Chuyện to nhỏ: Ông tập đoàn trưởng [Some issues: Production unit leader], *An Giang*, 13 June, p. 4.

An Giang. (1982). Chuyện to nhỏ: Ông cán bộ xã T [Some issues: Commune cadre], *An Giang*, 27 June, p. 4.

An Giang. (1982). Ban Nông Nghiệp Tỉnh Ủy An Giang: Thắng lợi của việc khoán sản phẩm trong nông nghiệp ở tỉnh nhà [An Giang Provincial Committee of Agriculture: The victory of the product contract in the province], *An Giang*, 4 July, p. 3.

An Giang. (1982). Trong tháng 6 phát triển 39 tập đoàn sản xuá, Tỉnh hiện có 474 tập đoàn [In June, 39 production units were established: The province now has 474 units], *An Giang*, 11 July, p. 1.

An Giang. (1982). Phú Tân tiến nhanh trong phong trào hợp tác xã hóa nông nghiệp [Collectivisation in Phú Tân advances fast], *An Giang*, 8 August, p. 2.

An Giang. (1982). Công tác điều chỉnh ruộng đất ở quê nhà [Land redistribution in rural areas], *An Giang*, 6 September, p. 4.

An Giang. (1982). Xã luận: Phát triển và củng cố tập đoàn sản xuất, hợp tác xã [The editorial: Improving and solidifying production units and collectives], *An Giang*, 3 October, p. 1.

An Giang. (1982). Chuyện to nhỏ: Xé rào [Some issues: Fence breaking], *An Giang*, 31 October, p. 4.

An Giang. (1983). Ban Tuyên Huấn tỉnh Ủy An Giang: Thành tích cải tạo nông nghiệp của tỉnh An Giang [An Giang Provincial Committee of Propaganda: The achievements of agricultural transformation in An Giang], *An Giang*, 2 January, p. 1.

An Giang. (1983). Các tập đoàn sản xuất, hợp tác xã tiến vào vụ Đông–Xuân 1982–1983 với nhiều khí thế mới [Production units and collectives entered into the winter–spring of 1982–1983 with new enthusiasm], *An Giang*, 2 January, p. 1.

An Giang. (1983). Chuyện to nhỏ: Nên chấm dứt [Some issues: Stop it], *An Giang*, 13 March, p. 4.

An Giang. (1983). Con số niềm tin [The figures and faith], *An Giang*, 20 March, p. 2.

An Giang. (1983). Lệ Làng [Village customs], *An Giang*, 24 April, p. 4.

An Giang. (1983). Xã luận: Tăng cường quản lý thị trường và ổn định giá cả [The editorial: Strengthening control of markets and stabilising prices], *An Giang*, 22 May, p. 1.

An Giang. (1983). Toàn tỉnh đẩy mạnh củng cố và phát triển tập đoàn [The province intensifies the solidification and extension of production units], *An Giang*, 12 June, p. 1.

An Giang. (1983). Các địa phương tập trung công tác củng cố, nâng chất và phát triển tập đoàn sản xuất [Local authorities must focus on solidifying, improving and extending production units], *An Giang*, 7 August, p. 2.

An Giang. (1983). Xã luận: Củng cố, nâng chất khâu cán bộ quản lý trong các hợp tác xã, tập đoàn sản xuất và tổ đoàn kết sản xuất [The editorial: Consolidating and improving the capacity of managerial cadres of collectives, production units, solidarity teams], *An Giang*, 7 August, p. 1.

An Giang. (1983). Chuyện to nhỏ: Đề nghị giải quyết thỏa đáng [Some issues: The need to solve the problem satisfactorily], *An Giang*, 28 August, p. 4.

An Giang. (1983). Đẩy mạnh cải tạo quan hệ sản xuất nông nghiệp [Speeding up agricultural transformation], *An Giang*, 25 September, p. 2.

An Giang. (1983). Văn Phòng Tỉnh Ủy An Giang: Tiếp tục điều chỉnh ruộng đất củng cố và phát triển tập đoàn sản xuất [An Giang Provincial Committee Office: Continuing land redistribution, and the solidification and extension of production units], *An Giang*, 9 October, p. 1.

An Giang. (1983). Hội nghị Tỉnh ủy để ra chương trình hành động từ nay đến năm 1984 [Provincial party committee meeting to make a plan of action from now to 1984], *An Giang*, 23 October, p. 1.

An Giang. (1983). Toàn tỉnh hiện có 1216 tập đoàn sản xuất, 57 liên tập đoàn, 70 tập đoàn máy nông nghiệp [An Giang now has 1,216 production units, 57 interproduction units and 70 machinery units], *An Giang*, 23 October, p. 2.

An Giang. (1983). Hợp tác xã Tây Huề qua 6 năm làm ăn tập thể [Tây Huề collective over the past 6 years], *An Giang*, 30 December, p. 2.

An Giang. (1984). Nguyễn Văn Nhung: Mít tinh trọng thể 10 năm giải phóng tỉnh An Giang và 40 năm Liên Xô chiến thắng phát xít Đức [Nguyễn Văn Nhung: A meeting to celebrate the 10-year anniversary of liberating An Giang province and 40 years of the Soviet Union victory over Nazi Germany], *An Giang*, 10 May, p. 1.

An Giang. (1984). Khắp nơi trong tỉnh [News around the province], *An Giang*, 12 July, p. 4.

An Giang. (1984). Tóa án Nhân dân tỉnh xét xử đầu cơ và hối lộ [Provincial People's Court adjudication on speculation and bribery], *An Giang*, 12 July, p. 3.

An Giang. (1984). Ban Nông Nghiệp Tỉnh Ủy: Tình hình điều chỉnh và qui hoạch ruộng đất ở xã Vĩnh Phú [Provincial Agriculture Board: Land redistribution in Vinh Phu Commune], *An Giang*, 9 August, p. 3.

An Giang. (1984). Xã luận: Đẩy mạnh cải tạo quan hệ sản xuất nông nghiệp [The editorial: Intensifying agricultural transformation], *An Giang*, 9 August, p. 1.

An Giang. (1984). Xã luận: Phải tập trung, củng cố, nâng chất các tập đoàn sản xuất [The editorial: The need to concentrate on improving and upgrading the quality of production units], *An Giang*, 18 August, p. 1.

An Giang. (1984). Chuyện to nhỏ: Khẩn trương nhưng vững chắc [Some issues: Hurry up and be firm in collectivisation], *An Giang*, 20 September, p. 4.

An Giang. (1984). Tô Sỹ Hồng: Một số nét chính trong cách quản lý ở các tập đoàn sản xuất và hợp tác xã nông nghiệp [Tô Sỹ Hồng: Some major issues in the management of production units and collectives], *An Giang*, 20 September, p. 1.

An Giang. (1984). Huyện Châu Phú xét xử bọn tham ô lương thực [Châu Thành District Court tries food thieves], *An Giang*, 27 September, p 3.

An Giang. (1984). Chuyện to nhỏ: Chuyện các ngài trong ban quản lý tập đoàn [Some issues: The problems caused by production unit managerial cadres], *An Giang*, 8 October, p. 4.

An Giang. (1984). Chuyện to nhỏ: Nợ không chụi trả [Some issues: Refusing to pay outstanding debt], *An Giang*, 16 November, p. 4.

An Giang. (1984). Phạm nhiều tội, 26 bị cáo ra tòa án nhân dân An Giang [26 defendants sentenced for many crimes in An Giang People's Court], *An Giang*, 27 December, p. 2.

An Giang. (1985). Ngành nông nghiệp tổng kết công tác năm 1984: Vượt qua khó khăn, toàn tỉnh gieo trồng 300,842 ha [Summing up 1984 agricultural production: Overcoming difficulties to cultivate 300,842 hectares], *An Giang*, 21 February, p. 1.

An Giang. (1985). Toàn tỉnh thành lập được 1957 tập đoàn sản xuất, tập thể hóa 106,798 ha [An Giang has 1,957 production units, collectivising 106,798 hectares], *An Giang*, 28 February, p. 1.

An Giang. (1985). Huyện Chợ Mới hoàn thành hợp tác hóa nông nghiệp [Chợ Mới district has completed collectivisation], *An Giang*, 4 April, p. 1.

An Giang. (1985). Qua hội nghị công báo hoàn thành cơ bản hợp tác hóa nông nghiệp ở Chợ Mới: Bài học gì được rút ra [Report from a conference announcing the completion of collectivisation in Chợ Mới: Lessons learned], *An Giang*, 15 April, p. 1.

An Giang. (1985). Trích diễn văn của đồng chí Lê Văn Nhung [An extract from Le Van Nhung's speech], *An Giang*, 10 May, p. 1.

An Giang. (1985). Chuyện to nhỏ [Some big and small issues], *An Giang*, 24 May, p. 4.

An Giang. (1985). Tại sao Châu Thành chưa giải quyết được tình trạng nợ trầm trọng? [Why haven't Châu Thành district authorities dealt with their huge outstanding debt?], *An Giang*, 31 May, p. 3.

An Giang. (1985). Đưa phong trào hợp tác hóa của tỉnh nhà lên vững chắc [Advancing collectivisation firmly], *An Giang*, 7 June, p. 1.

An Giang. (1985). Xã luận: Củng cố, nâng chất các tập đoàn một nhiệm vụ hết sức bức thiết [The editorial: Solidification and upgrading of production units are essential], *An Giang*, 12 July, p. 1.

An Giang. (1985). Toàn tỉnh đã xây dựng được 2570 tập đoàn sản xuất, 7 hợp tác xã và 21 liên tập đoàn sản xuất [The province has established 2,570 production units, 7 collectives and 21 interproduction units], *An Giang*, 2 August, p. 1.

An Giang. (1985). Trả lời bạn đọc: Về việc điều chỉnh ruộng đất ở xã Long Kiến [Reply to reader's letter: On land redistribution in Long Kiến Commune], *An Giang*, 27 September, p. 3.

An Giang. (1985). Trong quí III toàn tỉnh củng cố, nâng chất 393 tập đoàn sản xuất, tập thể hóa 310 máy cày [The province has upgraded 393 production units, and collectivised 310 ploughing machines in the third quarter of 1985], *An Giang*, 27 September, p. 1.

An Giang. (1985). Xã luận: Củng cố nâng chất 393 tập đoàn sản xuất, hợp tác xã nông nghiệp [The editorial: Strengthening the quality of 393 production units and collectives], *An Giang*, 27 September, p. 1.

An Giang. (1985). An Giang hoàn thành cơ bản công tác cải tạo nông nghiệp [An Giang has completed agricultural transformation], *An Giang*, 22 November, p. 1.

An Giang. (1985). Hoàn thành cơ bản công tác cải tạo nông nghiệp [The basic completion of agricultural transformation], *An Giang*, 22 November, p. 1.

An Giang. (1985). Phát biểu của đồng chí Võ Văn Bảo, phó bí thư tỉnh ủy tại hội nghị tổng kết hoàn thành cơ bản hợp tác hóa nông nghiệp tỉnh An Giang [Speech of Comrade Võ Văn Bảo, Deputy Secretary of Provincial Party Committee, at the meeting to complete basic agricultural cooperation in An Giang], *An Giang*, 22 November, pp. 1, 3.

An Giang. (1985). Chợ Mới vào vụ mới [Chợ Mới begins to cultivate a new crop], *An Giang*, 20 December, p. 2.

An Giang. (1986). Những khoảng cách trong sản xuất nông nghiệp ở Định Thành [The gaps between expectations and agricultural production in Định Thành], *An Giang*, 18 April, p. 2.

An Giang. (1986). Qua thanh tra có 45 tập đoàn, 5 liên tập đoàn sản xuất khoán trắng [Investigations found 45 production units and 5 interunits committed 'blank contracts'], *An Giang*, 18 April, p. 2.

An Giang. (1986). Hàng đổi hàng đến tay ai? [Who benefits from goods exchanged for paddy?], *An Giang*, 16 May, pp. 3, 4.

An Giang. (1986). Qua kiểm tra chất lượng ở một số tập đoàn [Evaluation of the quality of production units], *An Giang*, 6 June, p. 2.

An Giang. (1986). Huyện Chợ Mới tiến hành kiểm tra một số tập đoàn nông nghiệp [Chợ Mới district carried out investigations into some production units], *An Giang*, 13 June, p. 1.

An Giang. (1986). Các huyện Chợ Mới, Phú Tân, Châu Phú thực hiện phê bình trước quần chúng [Chợ Mới, Phú Tân and Châu Phú districts undertake public self-criticism], *An Giang*, 27 June, p. 1.

An Giang. (1986). Tự phê bình và phê bình: Ý kiến từ một cuộc họp [Criticism and self-criticism: Opinion from a meeting], *An Giang*, 4 July, p. 3.

An Giang. (1986). Nguyễn Vũ: Tiếp tục đưa nhịp độ phát triển nông nghiệp lên nhanh hơn [Nguyễn Vũ: Continue to speed up agricultural production], *An Giang*, 24 October, p. 1.

An Giang. (1986). Xã luận: Xây dựng cơ chế mới và chính sách phù hợp với các đơn vị sản xuất nông nghiệp [The editorial: Building new appropriate mechanisms to fit agricultural organisations], *An Giang*, 7 November, p. 1.

An Giang. (1986). Củng cố và cải tạo máy nông nghiệp, xay xát [Improving and renovating the management of agricultural machines], *An Giang*, 22 December, p. 2.

An Giang. (1986). Hội nghị cán bộ quán triệt nghị quyết 1987 [Officials' meeting thoroughly resolves 1987 resolution], *An Giang*, 22 December, p. 1.

An Giang. (1987). Phỏng vấn phó bí thư tỉnh ủy An Giang: Nhiệm kỳ tới sẽ cố gắng làm thế nào để góp phần vận dụng nghị quyết VI vào thực tế tỉnh nhà đạt kết quả cụ thể hơn nữa [Interview with Vice-Chairman of An Giang's Party Committee: The next term will try to contribute to the application of Resolution VI in the province to achieve more concrete results], *An Giang*, 17 March, p. 1.

An Giang. (1987). Tập đoàn trưởng trắng trợn ức hiếp tập đoàn viên [A production unit leader obviously bullied members], *An Giang*, 17 April, p. 8.

An Giang. (1987). Cần hiểu rõ và chấp hành tinh thần quyết định 93 của Ủy ban nhân dân tỉnh [The need to fully understand and abide by the Provincial People's Committee's Directive No. 93], *An Giang*, 22 May, p. 6.

An Giang. (1987). Mỗi tuần một chuyện: Nhanh chóng giải quyết vấn đề ruộng đất hợp lý [A story each week: Be quick to solve land disputes], *An Giang*, 22 May, p. 3.

An Giang. (1987). Mỗi tuần một chuyện: Hiểu lầm hay cố ý? [A story each week: Misunderstood or intended?], *An Giang*, 29 May, p. 7.

An Giang. (1987). Ý kiến: Không nên ngộ nhận giữa việc phân bổ chia cấp đất đai cho hợp lý với việc trả lại ruộng đất cho chủ cũ [The opinion piece: Don't mistake rational reallocation of land for returning land to previous landowners], *An Giang*, 29 May, p. 1.

An Giang. (1987). Xung quanh chuyện đầu tư cho sản xuất nông nghiệp [The problem of agricultural investment], *An Giang*, 17 July, p. 2.

An Giang. (1987). Còn thắc mắc về việc điều chỉnh ruộng đất ở xã Thạnh Mỹ Tây [Some concerns about land redistribution in Thạnh Mỹ Tây Commune], *An Giang*, 31 July, p. 6.

An Giang. (1987). Cai hợp đồng B [Boss of the B contracts], *An Giang*, 28 August, p. 7.

An Giang. (1987). Giá cả thu mua, chính sách thuế nông nghiệp ảnh hưởng đến đời sống của nông dân [Procurement prices and agricultural taxes affect peasants' living standards], *An Giang*, 28 August, p. 3.

An Giang. (1987). Mỗi tuần một chuyện: Chuyện ở tập đoàn sản xuất [A story each week: Production unit story], *An Giang*, 28 August, p. 7.

An Giang. (1987). Tình hình thanh lý nợ hợp đồng trong sản xuất nông nghiệp ở Thoại Sơn [Contract debt liquidation in agricultural production in Thoại Sơn district], *An Giang*, 4 September, p. 3.

An Giang. (1987). Phỏng vấn Nguyễn Vũ: Nhất định khắc phục những yếu kém đưa tập đoàn sản xuất tiến lên một bước [Interview with Nguyễn Vũ: Be certain in correcting shortcomings to advance production units], *An Giang*, 18 September, p. 2.

An Giang. (1987). Người nông dân đang cần phương thức đầu tư hợp lý phát triển sản xuất nông nghiệp [Peasants need a rational method of agricultural investment], *An Giang*, 23 October, p. 2.

An Giang. (1987). Phỏng vấn Nguyễn Hữu Khánh: Phải nhanh chóng xử lý tiêu cực ở những tập đoàn sản xuất vẫn còn do dự chưa giải quyết [Interview with Nguyễn Hữu Khánh: The need to quickly deal with the remaining negativism in production units], *An Giang*, 4 December, p. 2.

An Giang. (1988). Võ Quang Liêm: Vấn đề củng cố, nâng chất các tập đoàn sản xuất [Võ Quang Liêm: The matter of solidifying and upgrading production units], *An Giang*, 15 January, p. 1.

An Giang. (1988). Những điều nghe thấy từ thực tế [Some issues learned from reality], *An Giang*, 4 March, p. 3.

An Giang. (1988). Xác định lại mục đích cải tạo nông nghiệp [Redefining the objectives of agricultural transformation], *An Giang*, 4 March, p. 1.

An Giang. (1988). Nỗi oan trái của bà con nông dân Tân Lập [The grievances of peasants in Tân Lập], *An Giang*, 1 July, p. 3.

An Giang. (1988). Nỗi oan trái của bà con nông dân Tân Lập [The grievances of peasants in Tân Lập], *An Giang*, 8 July p. 3.

An Giang. (1988). Ý kiến: Làm chủ [The opinion piece: Being a master], *An Giang*, 29 July, p. 1.

An Giang. (1988). Những người bao chiếm đất [Land misusers], *An Giang*, 5 August, p. 2.

An Giang. (1988). Những người bao chiếm đất [Land misusers], *An Giang*, 19 August, p. 2.

An Giang. (1988). Báo cáo chính trị của Ban chấp hành đảng bộ tỉnh An Giang [The political report of the Executive Committee of An Giang's Provincial Party Committee], *An Giang*, 24 October, p. 3.

An Giang. (1988). Đất: Tiếng kêu từ phía nông dân [Land problem: A cry from peasants], *An Giang*, 18 November, p. 3.

An Giang. (1988). Phú Long: Cán bộ xã còn bao chiếm đất [Phú Long: Cadres still misappropriate land], *An Giang*, 9 December, p. 4.

Ban Bí Thư (BBT). (1976). *Chỉ thị 235-CT/TW của Ban bí thư Trung ương Đảng Cộng Sản Việt Nam (ngày 20 tháng 9 năm 1976) về việc thực hiện nghị quyết của Bộ chính trị về vấn đề ruộng đất ở Miền Nam* [*Directive No. 235 of the Secretariat of the Central Committee Communist Party of Vietnam (20 September 1976) on the Implementation of the Politburo's Land Resolution in the South*]. Hà Nội: Ban bí thư Trung ương Đảng Cộng Sản Việt Nam.

Ban Cải Tạo Nông Nghiệp An Giang (BCTNNAG). (1978). *Báo cáo tình hình cải tạo xã hội chủ nghĩa* [*Report on Socialist Agricultural Transformation*], 13 December. Long Xuyên: Ban Cải Tạo Nông Nghiệp An Giang.

Ban Cải Tạo Nông Nghiệp Miền Nam (BCTNNMN). (1978). *Thông báo về cuộc họp từ ngày 22–24 tháng 10 năm 1979 của Ban cải tạo nông nghiệp Miền Nam* [*Report of Central Committee for Agricultural Transformation in the South on 22–24 October 1979 Meeting*], 5 November. Hồ Chí Minh: Ban Cải Tạo Nông Nghiệp Miền Nam.

Ban Cải Tạo Nông Nghiệp Miền Nam (BCTNNMN). (1979). *Bài của đồng chí Võ Chí Công: Kết luận hội nghị cải tạo nông nghiệp các tỉnh B2 cũ* [*Speech by Võ Chí Công: At the Conference on Agricultural Transformation in the Old B2 Zone*], 26 August. Hồ Chí Minh: Ban Cải Tạo Nông Nghiệp Miền Nam.

Ban Cải Tạo Nông Nghiệp Miền Nam (BCTNNMN). (1979). *Thông tri về việc kịp thời và ra sức củng cố các tập đoàn sản xuất nông nghiệp* [*Announcement on Doing the Best to Improve Production Units*], 1 November. Hồ Chí Minh: Ban Cải Tạo Nông Nghiệp Miền Nam.

Ban Cải Tạo Nông Nghiệp Miền Nam (BCTNNMN). (1984). *Báo cáo tình hình ruộng đất và quá trình điều chỉnh ruộng đất trong nông thôn Nam Bộ* [*Report on Land Redistribution in the Southern Region*], January. Hồ Chí Minh: Ban Cải Tạo Nông Nghiệp Miền Nam.

Ban Cải Tạo Nông Nghiệp Minh Hải (BCTNNMH). (1979). *Dự thảo báo cáo: Nhận định, đánh giá tình hình cải tạo nông nghiệp thời gian qua ở Minh Hải* [*A Draft Report: Evaluation of Agricultural Transformation in Minh Hải*], 13 November. Minh Hải: Ban Cải Tạo Nông Nghiệp tỉnh Minh Hải.

Ban Chấp Hành Đảng Bộ Huyện Chợ Mới (BCHDBHCM). (1995). *Lịch sử Đảng bộ huyện Chợ Mới* [*The History of Chợ Mới Party Cell, 1927–1995*]. Chợ Mới: Ban Chấp Hành Đảng Bộ Huyện Chợ Mới.

Ban Chấp Hành Đảng Bộ Quảng Nam-Đà Nẵng. (1986). *Báo cáo tình hình và nhiệm vụ của Ban chấp hành đảng bộ tỉnh lần thứ 14 Đảng bộ tỉnh Quảng Nam-Đà Nẵng* [*Report of the Provincial Executive Committee of the 14th Provincial Party Committee of Quảng Nam-Đà Nẵng on the Economic Situation and Ongoing Tasks*], 4 October. Tam Kỳ: Quảng Nam-Đà Nẵng.

Ban Chấp Hành Trung Ương (BCHTU). (1977). *Chỉ thị 29-CT/TW về chính sách được áp dụng ở các hợp tác xã thí điểm ở Miền Nam* [*Directive No. 29-CT/TW on Policy for Pilot Collectives in the South*], 26 December. Hà Nội: NXB Nông Nghiệp.

Ban Chấp Hành Trung Ương (BCHTU). (1993). Chỉ thị cải tiến công tác khoán, mở rộng khoán sản phẩm đến nhóm lao động và người lao động trong hợp tác xã nông nghiệp (ngày 13 tháng 1 năm 1981) [Directive on improving the contracting of products to labour groups and labourers in agricultural cooperatives (13 January 1981)]. In Bộ Nông Nghiệp and Công Nghiệp Thực Phẩm (eds), *Chủ trương chính sách của Đảng, Nhà nước và tiếp tục đổi mới và phát triển nông nghiệp và nông thôn* [*Vietnam's Agrarian Policies*]. Hà Nội: NXB Nông Nghiệp.

Ban Kinh Tế Tỉnh Ủy QN-ĐN. (1985). *Tốc độ khôi phục kinh tế và phát triển xã hội của tỉnh gần 10 năm giải phóng* [*The Economic Performance of the Province over the Past 10 Years*], 16 February. Tam Kỳ: Quảng Nam-Đà Nẵng.

Ban Nông Nghiệp Tỉnh Ủy QN-ĐN. (1984). *Những vấn đề cần giải quyết để phát huy động lực của chế độ khoán mới trong hợp tác xã sản xuất nông nghiệp* [*Some Ideas to Facilitate the Incentives for Product Contracts*], 24 November. Tam Kỳ: Quảng Nam-Đà Nẵng.

Ban Quản Lý HTX NN TU. (1982). *Khoán sản phẩm trong hợp tác xã và tập đoàn sản xuất nông nghiệp* [*The Product Contract in Collectives and Production Units*]. Hà Nội: NXB Sự Thật.

Ban Thường Vụ Tỉnh Ủy QN-ĐN. (1984). *Nghị quyết 53/CT-TV về việc tiếp tục khuyến khích phát triển kinh tế gia đình* [*Provincial Resolution No. 53/CT-TV on Continually Facilitating the Household Economy*], 20 December. Tam Kỳ: Quảng Nam-Đà Nẵng.

Ban Tuyên Huấn Trung Ương. (1988). *Đảng trả lời nông dân một số vấn đề cấp bách về ruộng đất* [*The Party's Response to Urgent Land Problems*]. Hồ Chí Minh: NXB Tuyên Huấn.

BBC Vietnamese. (2016). Dân nhiều vùng lên Hà Nội biểu tình đòi đất [Many people in Hanoi protest about land], *BBC Vietnamese*, 21 January. Available from: www.bbc.com/vietnamese/forum/2016/01/160120_quynhchau_land_protests (accessed 4 October 2017).

Beresford, M. (1988). Issues in economic unification: Overcoming the legacy of separation. In D. Marr and C. White (eds), *Postwar Vietnam: Dilemmas in Socialist Development*. Ithaca, NY: Cornell University Press.

Bộ Nông Nghiệp. (1990). *Dự thảo tổng kết 3 năm thực hiện nghị quyết 10 của Bộ chính trị về đổi mới quản lý kinh tế nông nghiệp* [*A Draft Summing Up of the Three-Year Implementation of Resolution No. 10*], 10 December. Hà Nội: Bộ Nông Nghiệp.

Bray, F. (1994). *The Rice Economies: Technology and Development in Asian Societies*. Berkeley, CA: University of California Press.

Callison, C. S. (1983). *Land-to-the-Tiller in the Mekong Delta: Economic, Social, and Political Effects of Land Reform in Four Villages of South Vietnam*. New York: University Press of America.

Chayanov, A. V. (1986). *The Theory of Peasant Economy*. Madison: University of Wisconsin Press.

Chi Cục Thống Kê huyện Chợ Mới (CCTKCM). (1984). *Niên giám thống kê 1976–1984 huyện Chợ Mới tỉnh An Giang* [*Chợ Mới District, An Giang Province, Statistical Year Book, 1976–1984*]. Chợ Mới: Chi Cục Thống Kê huyện Chợ Mới.

Christodoulou, D. (1990). *The Unpromised Land: Agrarian Reform and Conflict Worldwide*. London: Zed Books.

Cục Thống Kê An Giang (CTKAG). (1986). *Tình hình kinh tế xã hội tỉnh An Giang 1983–1985* [*An Giang's Socioeconomic Situation from 1983–1985*]. Long Xuyên: Cục Thống Kê An Giang.

Cục Thống Kê An Giang (CTKAG). (2000). *Niên giám thống kê tỉnh An Giang* [*An Giang Statistical Year Book*]. Long Xuyên: Cục Thống Kê An Giang.

Cục Thống Kê An Giang (CTKAG). (2005). *Tổng hợp diện tích, năng suất sản lượng cây trồng hàng năm và số lượng gia súc gia cầm gia đoạn 1975–2005* [*Area, Productivity and Output of Annual Crops in An Giang from 1975–2005*]. Long Xuyên: Cục Thống Kê An Giang.

Cục Thống Kê tỉnh Quảng Nam (CTKQN). (2005). *Quảng Nam 30 Năm Xây Dựng và Phát triển* [*Quảng Nam's Socioeconomic Development over the Past 30 Years*]. Tam Kỳ: Cục Thống Kê tỉnh Quảng Nam.

Dahm, B., Houben, V. J. H., Grossheim, M., Endres, K. W. and Spitzenpfeil, A. (1999). *Vietnamese Villages in Transition: Background and Consequences of Reform Policies in Rural Vietnam*. Passau, Germany: Department of Southeast Asian Studies, University of Passau.

Đại Đoàn Kết. (1977). Nghị quyết lần thứ II: Ban chấp hành Trung ương Đảng khóa IV ra nghị quyết [Resolution II of the Central Committee of the Party IV], *Đại Đoàn Kết*, 3 September.

Đại Đoàn Kết. (1977). Nhiệm vụ cải tạo quan hệ sản xuất Miền Nam [Ongoing task for socialist transformation in the south], *Đại Đoàn Kết*, 17 September.

Đảng Bộ Chợ Mới (ĐBCM). (2000). Trên mặt trận bảo vệ an ninh tổ quốc [On the national security front]. In *Chợ Mới 25 năm xây dựng và phát triển* [*Chợ Mới's Socioeconomic Development over the Past 25 Years*]. Chợ Mới: Đảng Bộ huyện Chợ Mới.

Đảng bộ huyện Gò Công. (1978). Vận động thành lập hợp tác xã thí điểm ở Gò Công [Mobilising and establishing pilot collectives in Gò Công]. In Võ Chí Công, Nguyễn Thành Thơ, Phan Văn Đáng and Phạm Văn Kiết (eds), *Con đường làm ăn tập thể của nông dân* [*The Collective Farmer's Way*]. Hồ Chí Minh: NXB Tp. Hồ Chí Minh.

Đảng Cộng Sản Việt Nam (ĐCSVN). (1982). *Văn kiện Đại hội đại biểu toàn quốc lần thứ V* [*Document of the Fifth National Congress*]. Hà Nội: NXB Sự Thật.

Đảng Cộng Sản Việt Nam (ĐCSVN). (2004). Báo cáo chính trị của Ban chấp hành Trung ương Đảng tại Đại hội đại biểu toàn quốc lần thứ IV, do đồng chí Lê Duẩn trình bày [Political report of the Party Executive Committee at the fourth national representative meeting]. In ĐCSVN, *Văn Kiện Đảng Toàn Tập: Tập 37, 1976* [*Party Document: Volume 37, 1976*]. Hà Nội: NXB Chính Trị Quốc Gia.

Đảng Cộng Sản Việt Nam (ĐCSVN). (2004). Báo cáo của Bộ chính trị tại hội nghị lần thứ hai Ban chấp hành Trung ương khóa IV [Report of the Politburo at the Second Conference of the Central Committee, Session IV]. In ĐCSVN, *Văn Kiện Đảng Toàn Tập: Tập 38, 1977* [*Party Document: Volume 38, 1977*]. Hà Nội: NXB Chính Trị Quốc Gia.

Đảng Cộng Sản Việt Nam (ĐCSVN). (2004). Báo cáo của Bộ chính trị tại Hội nghị Trung Ương Đảng lần thứ 24 [Report of the Politburo at the 24th Party Central Committee Conference]. In ĐCSVN, *Văn Kiện Đảng Toàn Tập: Tập 36, 1975* [*Party Document: Volume 36, 1975*]. Hà Nội: NXB Chính Trị Quốc Gia.

Đảng Cộng Sản Việt Nam (ĐCSVN). (2004). Báo cáo tổng kết công tác xây dựng Đảng và sửa đổi điều lệ Đảng (ngày 17 tháng 12 năm 1976) [Report on building party organisation and changing party regulations (17 December 1976)]. In ĐCSVN, *Văn Kiện Đảng Toàn Tập: Tập 37, 1976* [*Party Document: Volume 37, 1976*]. Hà Nội: NXB Chính Trị Quốc Gia.

Đảng Cộng Sản Việt Nam (ĐCSVN). (2004). Chỉ thị 57/CT-TW về việc xóa bỏ các hình thức bóc lột của phú nông, tư sản nông thôn và tàn dư bóc lột phong kiến [Directive No. 57 on eliminating exploitation in the south]. In ĐCSVN, *Văn Kiện Đảng Toàn Tập, Tập 38, 1977* [*Party Document: Volume 38, 1977*]. Hà Nội: NXB Chính Trị Quốc Gia.

Đảng Cộng Sản Việt Nam (ĐCSVN). (2004). Chỉ thị của Ban bí thư, số 273/CT-TW (ngày 24 tháng 9 năm 1976) về việc củng cố tổ chức cở sở Đảng và kết nạp Đảng viên mới ở Miền Nam [Secretariat's Directive No. 273 (24 September 1976) on consolidating party organisation in the south]. In ĐCSVN, *Văn Kiện Đảng Toàn Tập: Tập 37, 1976* [*Party Document: Volume 37, 1976*]. Hà Nội: NXB Chính Trị Quốc Gia.

Đảng Cộng Sản Việt Nam (ĐCSVN). (2004). Chỉ thị của Bộ chính trị, số 43/CT-TW (ngày 14 tháng 4 năm 1978) về việc nắm vững và đẩy mạnh công tác cải tạo nông nghiệpở Miền Nam) [Politburo's Directive No. 43 (14 April 1978) on intensifying agricultural transformation in the south]. In ĐCSVN, *Văn Kiện Đảng Toàn Tập: Tập 39, 1978* [*Party Document: Volume 39, 1978*]. Hà Nội: NXB Chính Trị Quốc Gia.

Đảng Cộng Sản Việt Nam (ĐCSVN). (2004). Đề Cương kết luận của đồng chí Lê Duẩn tại Hội nghị lần thứ II [Lê Duẩn's final statements at second plenum]. In ĐCSVN, *Văn Kiện Đảng Toàn Tập: Tập 38, 1977* [*Party Document: Volume 38, 1977*]. Hà Nội: NXB Chính Trị Quốc Gia.

Đảng Cộng Sản Việt Nam (ĐCSVN). (2004). Nghị quyết của Bộ chính trị số 254/NQ-TW (ngày 15 tháng 7 năm 1976) về những công tác trước mắt ở Miền Nam [Politburo Resolution No. 254/NQ-TW (15 July 1976) on ongoing work in the south]. In ĐCSVN, *Văn Kiện Đảng Toàn Tập: Tập 37, 1976* [*Party Document: Volume 37, 1976*]. Hà Nội: NXB Chính Trị Quốc Gia.

Đảng Cộng Sản Việt Nam (ĐCSVN). (2004). Nghị quyết của Đại hội Đảng lần thứ IV của Đảng Cộng Sản Việt Nam (ngày 20 tháng 12 năm 1976) [Resolution of the Fourth Party Congress of the Communist Party of Vietnam (20 December 1976)]. In ĐCSVN, *Văn Kiện Đảng Toàn Tập: Tập 37, 1976* [*Party Document: Volume 37, 1976*]. Hà Nội: NXB Chính Trị Quốc Gia.

Đảng Cộng Sản Việt Nam (ĐCSVN). (2004). Nghị quyết hội nghị lần thứ 6 Ban chấp hành Trung ương Đảng khóa IV [Resolution of the 6th Plenum of the Fourth Party Central Committee]. In ĐCVSN, *Văn Kiện Đảng Toàn Tập: Tập 40, 1979 [Party Document: Volume 40, 1979]*. Hà Nội: NXB Chính Trị Quốc Gia.

Đảng Cộng Sản Việt Nam (ĐCSVN). (2004). Nghị quyết hội nghị lần thứ 24 của Ban Chấp Hành Trung Ương ĐCSVN, số 247/NQ-TW (ngày 29 tháng 9 năm 1975) [Resolution No. 247/NQ-TW (29 September 1975)]. In ĐCSVN, *Văn Kiện Đảng Toàn Tập: Tập 36, 1975 [Party Document: Volume 36, 1975]*. Hà Nội: NXB Chính Trị Quốc Gia.

Đảng Cộng Sản Việt Nam (ĐCSVN). (2004). Nghị quyết Hội nghị lần thứ hai của Ban chấp hành Trung ương Đảng khóa IV, số 03/NQ-TW (ngày 19 tháng 8 năm 1977) [Resolution No. 03/NQ-TW of the Second Plenum of the Central Committee of the Party IV (19 August 1977)]. In ĐCSVN, *Văn Kiện Đảng Toàn Tập: Tập 38, 1977 [Party Document: Volume 38, 1977]*. Hà Nội: NXB Chính Trị Quốc Gia.

Đảng Cộng Sản Việt Nam (ĐCSVN). (2004). Nghị quyết lần thứ 24 của Ban chấp hành Trung ương Đảng khóa III [Resolution No. 24 of the Third Party Central Committee]. In ĐCSVN, *Văn Kiện Đảng Toàn Tập: Tập 36, 1975 [Party Document: Volume 36, 1975]*. Hà Nội: NXB Chính Trị Quốc Gia.

Đảng Cộng Sản Việt Nam (ĐCSVN). (2004). Phương hướng nhiệm vụ và mục tiêu chủ yếu của kế hoạch 5 năm 1976–1980 [Key tasks and objectives of the five-year plan, 1976–1980]. In ĐCSVN, *Văn Kiện Đảng Toàn Tập: Tập 37, 1976 [Party Document: Volume 37, 1976]*. Hà Nội: NXB Chính Trị Quốc Gia.

Đảng Cộng Sản Việt Nam (ĐCSVN). (2005). Chỉ thị của Ban bí thư số 02/CT-TW (ngày 21 tháng 1 năm 1977) về những việc trước mắt để giải quyết lương thực [Directive of the Secretariat No. 02/CT-TW (21 January 1977) on immediate matters for food processing]. In ĐCSVN, *Văn Kiện Đảng Toàn Tập: Tập 38, 1977 [Party Document: Volume 38, 1977]*. Hà Nội: NXB Chính Trị Quốc Gia.

Đảng Cộng Sản Việt Nam (ĐCSVN). (2005). Chỉ thị của Ban bí thư số 15/CT-TW (ngày 4 tháng 8 năm 1977) về việc thí điểm cải tạo xã hội chủ nghĩa ở Miền Nam [Secretariat's Directive No. 15/CT-TW (4 August 1977) on experimenting with socialist agricultural transformation in the south]. In ĐCSVN, *Văn Kiện Đảng Toàn Tập: Tập 38, 1977* [*Party Document: Volume 38, 1977*]. Hà Nội: NXB Chính Trị Quốc Gia.

Đảng Cộng Sản Việt Nam (ĐCSVN). (2005). Chỉ thị của Ban Bí Thư số 19/CT-TW (ngày 3 tháng 5 năm 1983) [Directive of the Secretariat No. 19/CT-TW (3 May 1983)]. In ĐCSVN, *Văn Kiện Đảng Toàn Tập: Tập 44, 1983* [*Party Document: Volume 44, 1983*]. Hà Nội: NXB Chính Trị Quốc Gia.

Đảng Cộng Sản Việt Nam (ĐCSVN). (2005). Chỉ thị của Ban bí thư số 93/CT-TW (ngày 30 tháng 6 năm 1980) [Directive of the Secretariat No. 93/CT-TW (30 June 1980)]. In ĐCSVN, *Văn Kiện Đảng Toàn Tập: Tập 41, 1980* [*Party Document: Volume 41, 1980*]. Hà Nội: NXB Chí Trị Quốc Gia.

Đảng Cộng Sản Việt Nam (ĐCSVN). (2005). Thông báo 14/TB-TW, ngày 20 tháng 4 năm 1981: Kết luận của Ban bí thư tại Hội nghị bàn việc xúc tiến công tác cải tạo nông nghiệp ở các tỉnh Nam Bộ [Circular No. 14/TB-TW, 20 April 1981: On facilitating agricultural transformation in the Southern Region]. In ĐCSVN, *Văn Kiện Đảng Toàn Tập: Tập 42, 1981* [*Party Document: Volume 42, 1981*]. Hà Nội: NXB Chính Trị Quốc Gia.

Đảng Cộng Sản Việt Nam (ĐCSVN). (2005). Thông tri của Ban bí thư số 138/TT-TW ngày 11 tháng 11 năm 1981 [Secretariat Circular No. 138/TT-TW of 11 November 1981]. In ĐCSVN, *Văn Kiện Đảng Toàn Tập: Tập 42, 1981* [*Party Document: Volume 42, 1981*]. Hà Nội: NXB Chính Trị Quốc Gia.

Đặng Phong. (2009). *Tư Duy Kinh Tế Việt Nam 1975–1989* [*The Economics of Vietnam 1975–1989*]. Hà Nội: Nhà Xuất Bản Trí Thức.

Đào Duy Huấn. (1988). Củng cố và hoàn thiện quan hệ sản xuất xã hội chủ nghĩa trong nông nghiệp tập thể hiện nay ở vùng Đồng Bằng Sông Cửu Long [Solidifying and perfecting socialist production relations in the agriculture of the Mekong Delta]. PhD thesis, Học Viện Nguyễn Ái Quốc, Hà Nội.

Elliott, D. W. (2003). *The Vietnamese War: Revolution and Social Change in the Mekong Delta 1930–1975*. Armonk, NY: M. E. Sharpe.

Fforde, A. and de Vylder, S. (1996). *From Plan to Market: The Economic Transition in Vietnam*. Boulder, CO: Westview Press.

Giao, H. (1984). Bước đi và hình thức hợp tác hóa nông nghiệp [Steps and forms of agricultural collaboration]. In Trần Xuân Bách, Nguyễn Ngọc Trìu and Hồng Giao (eds), *Bước Đi và Hình Thức Hợp Tác Hóa Nông Nghiệp* [*Steps and Forms of Agricultural Cooperation*]. Hồ Chí Minh: NXB Tổng Hợp TP.

Grossheim, M. (1999). The impact of reforms on the agricultural sector in Vietnam: The land issue. In B. Dahm, V. J. H. Houben, M. Grossheim, K. W. Endres and A. Spitzenpfeil (eds), *Vietnamese Villages in Transition: Background and Consequences of Reform Policies in Rural Vietnam*. Passau, Germany: Department of Southeast Asian Studies, University of Passau.

Hardin, G. (1968). The tragedy of the commons. *Science* 162(3859): 1243–8. doi.org/10.1126/science.162.3859.1243.

Hayami, Y. and Godo, Y. (2005). *Development Economics: From the Poverty to the Wealth of Nations*. Oxford: Oxford University Press. doi.org/10.1093/0199272700.001.0001.

Hicks, N. (2005). Organizational adventures in district government. PhD thesis, The Australian National University, Canberra.

Hội Đồng Chính Phủ (HĐCP). (1976). *Quyết định số 188/CP của Hội Đồng Chính Phủ (ngày 25 tháng 9 năm 1976) về chính sách xóa bỏ tàn tích chiếm hữu ruộng đất và các hình thức bóc lột thực dân, phong kiến ở Miền Nam Việt Nam* [*Ministerial Council's Decision No. 188/ CP (25 September 1976) on the Policy of Eliminating Land Tenure and Other Forms of Colonial and Feudal Exploitation in the South*]. Hà Nội: Hội Đồng Chính Phủ.

Hồng Giao. (1984). *Đưa Nông nghiệp lên một bước lớn Xã hội chủ nghĩa* [*Taking Agriculture One Step Towards Socialist Large-Scale Production*]. Hà Nội: NXB Sự Thật.

HTX Nghĩa Lâm. (1978). Kinh nghiệm xây dựng hợp tác xã Nghĩa Lâm, tỉnh Nghĩa Bình [Experiences from establishing Nghĩa Lâm collective in Nghĩa Bình province]. In Võ Chí Công, Nguyễn Thành Thơ, Phan Văn Đáng and Phạm Văn Kiết (eds), *Con đường làm ăn tập thể của nông dân* [*The Collective Farmer's Way*]. Hồ Chí Minh: NXB Tp. Hồ Chí Minh.

Huy Đức. (2012). *Bên Thắng Cuộc* [*The Winning Side*]. 2 vols. Giai Phong: OsinBook.

Huỳnh Thị Gấm. (1998). Những biến đổi kinh tế xã hội ở nông thôn Đồng bằng sông Cửu Long 1975–1995 [Socioeconomic changes in the Mekong Delta from 1975–1995]. PhD thesis, Đại Học Khoa Học Xã Hội Nhân Văn, Hồ Chí Minh.

Kelliher, D. (1992). *Peasant Power in China: The Era of Rural Reform, 1979–1989*. New Haven, CT: Yale University Press.

Kerkvliet, B. (1999). Accelerating cooperatives in rural Vietnam, 1955–1961. In B. Dahm, V. J. H. Houben, M. Grossheim, K. W. Endres and A. Spitzenpfeil (eds), *Vietnamese Villages in Transition: Background and Consequences of Reform Policies in Rural Vietnam*. Passau, Germany: Department of Southeast Asian Studies, University of Passau.

Kerkvliet, B. J. (2005). *The Power of Everyday Politics: How Vietnamese Peasants Transformed National Policy*. Ithaca, NY: Cornell University Press.

Kerkvliet, B. J. T. (1993). Claiming the land: Takeovers by villagers in the Philippines with comparisons to Indonesia, Peru, Portugal, and Russia. *The Journal of Peasant Studies* 20(3): 459–93. doi.org/10.1080/03066159308438518.

Kerkvliet, B. J. T. (1995). Village–state relations in Vietnam: The effect of everyday politics. *Journal of Asian Studies* 54(2): 396–418. doi.org/10.2307/2058744.

Kurtz, M. J. (2000). Understanding peasant revolution: From concept to theory and case. *Theory and Society* 29(1): 93–124. doi.org/10.1023/A:1007059213368.

Lâm Quang Huyên. (1985). *Cách mạng ruộng đất ở Miền Nam Việt Nam* [*The Land Revolution in South Vietnam*]. Hà Nội: NXB Khoa Học Xã Hội.

Lâm Quang Huyên. (2004). *Kinh tế nông hộ và kinh tế hợp tác trong nông nghiệp Việt Nam* [*The Peasant Household Economy and the Collective Economy in Vietnam*]. Hồ Chí Minh: NXB Trẻ.

Lê Duẩn. (1980). *Cải Tạo Xã Hội Chủ Nghĩa ở Miền Nam* [*Socialist Transformation in the South*]. Hà Nội: NXB Sự Thật.

Lê Duẩn. (1985). Báo cáo chính trị tại đại hội đại biểu toàn quốc lần thứ V của Đảng [Political report at Fifth National Party Congress]. In *Cách Mạng Xã Hội Chủ Nghĩa ở Việt Nam (Tác hầm chọn lọc), Tập IV* [*Social Revolution in Vietnam (Selected Activities). Volume 4*]. Hà Nội: NXB Sự Thật.

Lê Duẩn. (2004). Toàn dân đoàn kết xây dựng tổ quốc Việt Nam thống nhất, xã hội chủ nghĩa [Calling for the whole country's solidarity to build a socialist and unified country]. In ĐCSVN, *Văn Kiện Đảng Toàn Tập: Tập 37, 1976* [*Party Document: Volume 37, 1976*]. Hà Nội: NXB Chính Trị Quốc Gia.

Lê Thanh Nghị. (1981). *Cải tiến công tác khoán, mở rộng khoán để thúc đẩy sản xuất, củng cố HTX nông nghiệp* [*Improving the Product Contract to Solidify Collectives*]. Hà Nội: NXB Sự Thật.

Lê Thị Lộc Mai. (2001). Quá trình giải quyết vấn đề ruộng đất và phát triển nông thôn ở Vĩnh Long giai đoạn Đổi mới 1986–1996 [Dealing with land problems to facilitate rural development in Vinh Long in the period 1986–1996]. Masters thesis, Đại học Khoa học Xã hội and Nhân văn, Hồ Chí Minh.

L. K. (1990). Từ quá khứ đến hiện tại: Mười lăm năm ấy [From past to present: Over the past 10 years]. Unpublished essay.

Lương Hồng Quang. (1996). *Văn hóa cộng đồng làng vùng Đồng Bằng Sông Cửu Long thập kỷ 80–90* [*Village Cultures in the Mekong Delta from the 1980s–1990s*]. Hà Nội: NXB Viện Văn Hóa.

Luong, H. V. (1992). *Revolution in the Village: Tradition and Transformation in North Vietnam, 1925–1988*. Honolulu: University of Hawai'i Press.

Malarney, S. K. (1998). State stigma, family prestige, and the development of commerce in the Red River Delta of Vietnam. In R. W. Hefner (ed.), *Market Cultures: Society and Morality in the New Asian Capitalisms*. Boulder, CO: Westview Press.

Migdal, J. S. (1988). *Strong Societies and Weak States: State–Society Relations and State Capabilities in the Third World*. Princeton, NJ: Princeton University Press.

Migdal, J. S. (2001). *State in Society: Studying How States and Societies Transform and Constitute One Another*. Cambridge: Cambridge University Press. doi.org/10.1017/CBO9780511613067.

Moise, E. (1982). The moral economy dispute. *Bulletin of Concerned Asian Scholars* 14(1): 72–7. doi.org/10.1080/14672715.1982.1041 2639.

Moise, E. E. (1976). Land reform and land reform errors in North Vietnam. *Pacific Affairs* 49(1): 70–92. doi.org/10.2307/2756362.

Moise, E. E. (1983). *Land Reform in China and North Vietnam: Consolidating the Revolution at the Village Level*. Chapel Hill, NC: University of North Carolina Press.

Ngo Vinh Long. (1988). Some aspects of cooperativization in the Mekong Delta. In D. Marr and C. White (eds), *Postwar Vietnam: Dilemmas in Socialist Development*. Ithaca, NY: Cornell University Press.

Nguyễn Đình Đầu. (1992). *Chế Độ Công Điền Công Thổ Trong Lịch Sử Khẩn Hoang Lập Ấp ở Nam Kỳ Lục Tỉnh* [Land Tenure System in the Southern Region of Vietnam in the History of Land Reclamation]. Hồ Chí Minh: NXB Trẻ.

Nguyễn Đức Bình. (1983). Tiếp tục suy nghĩ về khoán sản phẩm trong nông nghiệp [Some ideas about the product contract]. In *Khoán sản phẩm chế độ quản lý mới trong nông nghiệp* [*The Product Contract and New Management Methods in Agriculture*]. Hà Nội: NXB Sự Thật.

Nguyễn Dương Đáng. (1983). *Kinh tế nông nghiệp Xã hội chủ nghĩa* [*Economics of Socialist Agriculture*]. Hà Nội: NXB Nông Nghiệp.

Nguyễn Huy. (1985). *Mấy vấn đề lý luận và thực tiễn của cách mạng quan hệ trong nông nghiệp nước ta* [*Theories and Practices of Revolution in the Production Relations of Our Country's Agriculture*]. Hà Nội: NXB Khoa Học Xã Hội.

Nguyễn Khắc Viện. (1990). *15 năm ấy: 1975–1990* [*15 Years: 1975–1990*]. Hồ Chí Minh: NXB TP.

Nguyễn Minh Nhị. (2004). *An Giang: Lịch sử tháo gỡ đột phá và chủ động hội nhập kinh tế thế giới* [*An Giang: The History of Breakthroughs and Active Integration into the World Economy*], 15 August. Long Xuyên: Sở Nông Nghiệp và Phát Triển Nông Thôn An Giang.

Nguyễn Sinh Cúc. (1991). *Thực Trạng Nông Nghiệp, Nông Thôn và Nông Dân Việt Nam 1976–1990* [*Agricultural and Rural Development in Vietnam 1976–1990*]. Hà Nội: NXB Thống Kê.

Nguyễn Thành Nam. (2000). Việc giải quyết vấn đề ruộng đất trong quá trình đi lên sản xuất lớn ở Đồng bằng Sông Cửu Long 1975–1993 [Resolving land issues in the process of large-scale production in the Mekong Delta, 1975–1993]. PhD thesis, Đại Học Khoa Học Xã Hội and Nhân Văn, Hồ Chí Minh.

Nguyễn Thành Thơ. (1978). Ra sức tiến hành hợp tác hóa nông nghiệp [Do our best to implement collectivisation]. In Võ Chí Công, Nguyyễn Thành Thơ, Phan Văn Đáng and Phạm Văn Kiết (eds), *Con đường làm ăn tập thể của nông dân* [*The Collective Farmer's Way*]. Hồ Chí Minh: NXB Tp. Hồ Chí Minh.

Nguyễn Trần Trọng. (1980). *Những vấn đề công tác cải tạo và xây dựng nông nghiệp ở các tỉnh phía Nam* [*Ongoing Tasks for Transforming and Building the South's Agriculture*]. Hà Nội: NXB Nông Nghiệp.

Nhà xuất bản Bản Đồ. (2005). *Vietnam's Administrative Atlas*. Hà Nội: NXB Bản Đồ.

Nhân Dân. (1977). Hồ Nghinh: Quảng Nam-Đà Nẵng vượt bậc phát triển sản xuất nông nghiệp [Hồ Nghinh: Quảng Nam-Đà Nẵng has made great progress in agriculture], *Nhân Dân*, 8 March, p. 5.

Nhân Dân. (1980). Năm năm cải tạo xã hội chủ nghĩa đối với nông nghiệp ở Miền Nam [Five years of socialist reform for agriculture in the south], *Nhân Dân*, 29 April, p. 1.

Nolan, P. (1976). Collectivization in China: Some comparisons with the USSR. *The Journal of Peasant Studies* 3(2): 192–220. doi.org/10.1080/03066157608437978.

O'Rourke, D. (2004). *Community-Driven Regulation: Balancing Development and the Environment in Vietnam*. Cambridge, MA: MIT Press.

Ostrom, E. (2005). *Understanding Institutional Diversity*. Princeton, NJ: Princeton University Press.

Phạm Văn Chiến. (2003). *Lịch sử kinh tế Việt Nam* [*History of the Vietnamese Economy*]. Hà Nội: NXB Đại Học Quốc Gia.

Phạm Văn Kiết. (1978). Nông dân đang sôi nổi đi lên làm ăn tập thể [Peasants are eager for collective farming]. In Võ Chí Công, Nguyễn Thành Thơ, Phan Văn Đáng and Phạm Văn Kiết (eds), *Con đường làm ăn tập thể của nông dân* [*The Collective Farmer's Way*]. Hồ Chí Minh: NXB TP Hồ Chí Minh.

Phan Quang. (1981). *Đồng Bằng Sông Cửu Long* [*The Mekong Delta*]. Hà Nội: NXB Văn Hóa.

Phan Văn Đáng. (1978). Tập dượt đi lên hợp tác xã nông nghiệp' [Experiment with agricultural collectives]. In Võ Chí Công, Nguyễn Thành Thơ, Phan Văn Đáng and Phạm Văn Kiết (eds), *Con đường làm ăn tập thể của nông dân* [*The Collective Farmer's Way*]. Hồ Chí Minh: NXB Tổng Hợp TP Hồ Chí Minh.

Pierson, P. (2004). *History, Institutions and Social Analysis*. Princeton, NJ: Princeton University Press.

Popkin, S. L. (1979). *The Rational Peasant: The Political Economy of Rural Society in Vietnam*. Berkeley, CA: University of California Press.

Prosterman, R. L. and Riedinger, J. M. (1987). *Land Reform and Democratic Development*. Baltimore: Johns Hopkins University Press.

Quảng Đà. (1974). Lời kêu gọi ra sức gia tăng sản xuất, thực hành tiết kiệm [Do the best to increase food production and be thrifty], *Quảng Đà*, 30 April.

Quảng Đà. (1974). Quyết thắng trên mặt trận nông nghiệp [Be determined to win on the agricultural front], *Quảng Đà*, 30 April.

Quảng Đà. (1974). Ngành nông nghiệp tỉnh Quảng Đà tích cực chăm lo vụ mùa tháng 8 [Agricultural sector in Quảng Đà is positive about caring for August crops], *Quảng Đà*, 20 June, p. 1.

Quảng Nam-Đà Nẵng. (1975). Tổ đổi công vần công ở Sông Bình [Labour exchange teams in Sông Bình], *Quảng Nam-Đà Nẵng*, 8 May, p. 2.

Quảng Nam-Đà Nẵng. (1975). Đất này về với chúng ta [This land comes back to us], *Quảng Nam-Đà Nẵng*, 1 September.

Quảng Nam-Đà Nẵng. (1975). Ngày mai rừng mía bạt ngàn [Sugarcane crop will be extensive in the future], *Quảng Nam-Đà Nẵng*, 20 November.

Quảng Nam-Đà Nẵng. (1975). Đồng quế vắng bóng trâu cày, vườn hoang nhà trống dân gầy xác xơ [Fields in rural areas lack draught animals; gardens were abandoned, houses were empty, and the people were prostrate and hungry], *Quảng Nam-Đà Nẵng*, 15 December.

Quảng Nam-Đà Nẵng. (1976). Đẩy mạnh sản xuất và thực hành tiết kiệm giải quyết vấn đề lương thực cấp bách trước mắt [Increase production and be thrifty to immediately deal with urgent food shortage], *Quảng Nam-Đà Nẵng*, 16 February.

Quảng Nam-Đà Nẵng. (1976). Tăng vụ sản xuất xuân hè [New additional spring–summer crops], *Quảng Nam-Đà Nẵng*, 16 February, p. 4.

Quảng Nam-Đà Nẵng (1976). Đẩy mạnh công tác thủy lợi nhỏ để phục vụ sản xuất xuân hè và hè thu [Extending irrigation for the spring–summer and summer–autumn crops], *Quảng Nam-Đà Nẵng*, 8 March, p. 1.

Quảng Nam-Đà Nẵng. (1976). Nhân dân tỉnh ta chẳng những đánh giặc giỏi mà còn giàu nghị lực và tài năng sáng tạo trong xây dựng lại quê hương giàu đẹp [Our province's people fought the enemy and are building the country well], *Quảng Nam-Đà Nẵng*, 29 March.

Quảng Nam-Đà Nẵng. (1976). Duy An lập khu nghĩa địa mới [Duy An has established new graveyards], *Quảng Nam-Đà Nẵng*, 5 April, p. 2.

Quảng Nam-Đà Nẵng. (1976). Những mùa lúa đầu tiên [The first rice crops], *Quảng Nam-Đà Nẵng*, 19 April, p. 2.

Quảng Nam-Đà Nẵng. (1976). Toàn tỉnh sôi nổi ra quân làm thủy lợi lợi [People in the province are extending irrigation], *Quảng Nam-Đà Nẵng*, 12 May, p. 1.

Quảng Nam-Đà Nẵng. (1976). Phấn đấu mở rộng nhanh diện tích canh tác [Strive to extend cultivated area], *Quảng Nam-Đà Nẵng*, 26 June.

Quảng Nam-Đà Nẵng. (1976). Hoàn thành thắng lợi công việc chia cấp ruộng đất cho nông dân [Land redistribution among peasants successfully completed], *Quảng Nam-Đà Nẵng*, 7 August, p. 1.

Quảng Nam-Đà Nẵng. (1976). Vụ sản xuất xuân hè thắng lợi [The spring–summer crops have a good result], *Quảng Nam-Đà Nẵng*, 7 August, p. 1.

Quảng Nam-Đà Nẵng. (1976). Điện Bàn: Cả huyện là một công trường [Điện Bàn: The whole district is a working field], *Quảng Nam-Đà Nẵng*, 11 August.

Quảng Nam-Đà Nẵng. (1976). Ủy ban nhân dân ra chỉ thị về công tác quy hoạch mồ mả và nhà của của nhân dân [The Provincial People's Committee issued a directive to reallocate tombs and houses], *Quảng Nam-Đà Nẵng*, 28 August, p. 1.

Quảng Nam-Đà Nẵng. (1976). Hoàn thành thắng lợi vẻ vang nhiệm vụ xây dựng đất nước, xây dựng chế độ mới, con người mới Xã hội chủ nghĩa [Completing the task of building the country, the new regime and new socialist men], *Quảng Nam-Đà Nẵng*, 8 September.

Quảng Nam-Đà Nẵng. (1976). Nêu cao tinh thân tự lực tự cường trong sản xuất và xây dựng quê hương [Be self-reliant in ensuring food production and building the country], *Quảng Nam-Đà Nẵng*, 29 September.

Quảng Nam-Đà Nẵng. (1976). Tỉnh ta có khả năng tự giải quyết lương thực hay không? [Is our province able to solve our own food problem?], *Quảng Nam-Đà Nẵng*, 22 November, p. 1.

Quảng Nam-Đà Nẵng. (1976). Nhìn lại diện tích đất đai để thấy rõ khả năng tự giải quyết lương thực [Re-examining agricultural areas to evaluate our capacity for dealing with food problems], *Quảng Nam-Đà Nẵng*, 26 November, p. 1.

Quảng Nam-Đà Nẵng. (1976). Nước và sản xuất lương thực ở tỉnh ta [Irrigation and food production in our province], *Quảng Nam-Đà Nẵng*, 18 December, p. 1.

Quảng Nam-Đà Nẵng. (1976). Vì sao tỉnh ta đặt vấn đề giải quyết lương? [Why do we pay great attention to solving the food production problem?], *Quảng Nam-Đà Nẵng*, 22 December, p. 1.

Quảng Nam-Đà Nẵng. (1977). Nghị quyết hội nghị Ban chấp hành Đảng bộ tỉnh khóa 11 [Resolution of 11th Provincial Party Executive Committee], *Quảng Nam-Đà Nẵng*, 12 March, p. 1.

Quảng Nam-Đà Nẵng. (1977). Gióng đường cày thắng lợi [Be victorious in agriculture], *Quảng Nam-Đà Nẵng*, 26 April, p. 1.

Quảng Nam-Đà Nẵng. (1977). Toàn xã Thăng Phước làm ăn trong các tổ đổi công thường xuyên [The whole population of Thăng Phước Commune is organised into regular labour exchange teams], *Quảng Nam-Đà Nẵng*, 23 May.

Quảng Nam-Đà Nẵng. (1977). Hội nghị tổ đổi công toàn tỉnh thành công tốt đẹp [The conference on labour exchange teams achieved good results], *Quảng Nam-Đà Nẵng*, 25 June.

Quảng Nam-Đà Nẵng. (1977). Bản hướng dẫn nội dung xây dựng tổ đổi công có định mức, khoán việc [Guidelines for establishing production teams working according to norms and contracts], *Quảng Nam-Đà Nẵng*, 29 June, p. 1.

Quảng Nam-Đà Nẵng. (1977). Nghị Quyết hội nghị Ban chấp hành đảng bộ tỉnh (khóa 11) về vấn đề phát triển và cải tạo nông nghiệp [Resolution of the Eleventh Provincial Party Congress on agricultural transformation and improvement], *Quảng Nam-Đà Nẵng*, 7 September, p. 1.

Quảng Nam-Đà Nẵng. (1977). Phấn khởi nghiên cứu học tập Nghị quyết Tỉnh ủy về phát triển và cải tạo nông nghiệp [Studying the Provincial Party Committee's resolution on agricultural transformation], *Quảng Nam-Đà Nẵng*, 10 September, p. 1.

Quảng Nam-Đà Nẵng. (1977). Nghiêm cấm thương nhân mua trâu bò để giết thịt [Prohibiting private merchants from purchasing and slaughtering livestock], *Quảng Nam-Đà Nẵng*, 24 September, p. 1.

Quảng Nam-Đà Nẵng. (1977). Xã Duy Phước trước bước ngoặc lịch sử [Duy Phước commune and its historic turning-point], *Quảng Nam-Đà Nẵng*, 24 September, p. 1.

Quảng Nam-Đà Nẵng. (1977). 96 phần trăm hộ nông dân ở Bình Lãnh tự nguyện ký đơn vào hợp tác xã [96 per cent of households in Bình Lãnh voluntarily signed forms to join the collective], *Quảng Nam-Đà Nẵng*, 4 October, p. 1.

Quảng Nam-Đà Nẵng. (1977). Duy Phước, Hòa Tiến, Bình Lãnh đi vào con đường làm ăn tập thể [Duy Phước, Hòa Tiến, and Bình Lãnh entered into collective farming], *Quảng Nam-Đà Nẵng*, 4 October, p. 1.

Quảng Nam-Đà Nẵng. (1977). Hòa Tiến: 1,057 hộ tự nguyện đưa 379 ha ruộng đất vào làm ăn tập thể [Hòa Tiến: 1,057 households voluntarily put 379 hectares into collective farming], *Quảng Nam-Đà Nẵng*, 4 October.

Quảng Nam-Đà Nẵng. (1977). Thành lập Ban cải tạo nông nghiệp [Establishing a committee for agricultural transformation], *Quảng Nam-Đà Nẵng*, 4 October, p. 1.

Quảng Nam-Đà Nẵng. (1977). Bà con nông dân trong tỉnh hãy theo con đường làm ăn tập thể của Duy Phước, Bình Lãnh, Hòa Tiến [Peasants in the province should follow the collective farming paths of Duy Phước, Bình Lãnh and Hòa Tiến people], *Quảng Nam-Đà Nẵng*, 11 October, p. 1.

Quảng Nam-Đà Nẵng. (1977). Xây dựng các tổ sản xuất có định mức khoán việc [Establishing production teams working according to norms and contracts], *Quảng Nam-Đà Nẵng*, 22 October, p. 3.

Quảng Nam-Đà Nẵng. (1977). Đại Lộc xây dựng các tổ sản xuất có định mức khoán việc [Đại Lộc is establishing production teams working according to norms and contracts], *Quảng Nam-Đà Nẵng*, 26 October, p. 2.

Quảng Nam-Đà Nẵng. (1977). Mở đại hội xã viên thành lập hợp tác xã nông nghiệp Duy Phước [Members' congress held to establish Duy Phước collective], *Quảng Nam-Đà Nẵng*, 29 October, p. 1.

Quảng Nam-Đà Nẵng. (1977). Mở đại hội xã viên thành lập hợp tác xã nông nghiệp Bình Lãnh và Hòa Tiến [Members' congress held to establish Bình Lãnh and Hòa Tiến collectives], *Quảng Nam-Đà Nẵng*, 5 November, p. 1.

Quảng Nam-Đà Nẵng. (1977). Thăng Bình từ thủy lợi đi lên [Thăng Bình advanced by irrigation], *Quảng Nam-Đà Nẵng*, 30 November.

Quảng Nam-Đà Nẵng. (1978). Mùa xuân và mùa đông ở hợp tác xã Bình Lãnh [The achievements and challenges of Bình Lãnh collective], *Quảng Nam-Đà Nẵng*, 11 February, p. 2.

Quảng Nam-Đà Nẵng. (1978). Tỉnh ủy mở hội nghị bàn về cải tạo xã hội chủ nghĩa đối với nông nghiệp [Provincial Party Committee opens conference on socialist agricultural transformation], *Quảng Nam-Đà Nẵng*, 22 February, p. 1.

Quảng Nam-Đà Nẵng. (1978). Hướng dẫn phân phối thu nhập vụ Đông–Xuân [Income distribution guidelines for the winter–spring crops], *Quảng Nam-Đà Nẵng*, 21 April, p. 1.

Quảng Nam-Đà Nẵng. (1978). Bình An xây dựng tổ sản xuất có định mức khoán việc [Bình An is establishing production teams working according to norms and contracts], *Quảng Nam-Đà Nẵng*, 24 April, p. 3.

Quảng Nam-Đà Nẵng. (1978). Hợp tác xã Bình Lãnh vượt khó khăn giành thắng lợi bước đầu [Bình Lãnh collective overcame difficulties and gained first good results], *Quảng Nam-Đà Nẵng*, 13 May, p. 2.

Quảng Nam-Đà Nẵng. (1978). Thắng lợi bước đầu của phong trào Hợp tác hóa nông nghiệp [The first victory steps of collectivisation], *Quảng Nam-Đà Nẵng*, 13 May, p. 1.

Quảng Nam-Đà Nẵng. (1978). Hồ Nghinh: Thắng lợi của việc xây dựng thí điểm hợp tác là thắng lợi có ý nghĩa của toàn đảng bộ và nhân dân toàn tỉnh [Hồ Nghinh: The success of pilot collectives is a significant victory for the province's party and people], *Quảng Nam-Đà Nẵng*, 27 May, p. 1.

Quảng Nam-Đà Nẵng. (1978). Tổng kết xây dựng thí điểm hợp tác xã nông nghiệp [Summing up establishing pilot collectives], *Quảng Nam-Đà Nẵng*, 27 May, p. 1.

Quảng Nam-Đà Nẵng. (1978). Tích cực chuẩn bị mở rộng phong trào tổ chức hợp tác xã sản xuất nông nghiệp [Be positive towards the extension of collectivisation], *Quảng Nam-Đà Nẵng*, 10 June, p. 1.

Quảng Nam-Đà Nẵng. (1978). Chi bộ Bình Lãnh lãnh đạo xây dựng hợp tác xã nông nghiệp [Bình Lãnh party cell leads building of the agricultural collective], *Quảng Nam-Đà Nẵng*, 14 June, p. 2.

Quảng Nam-Đà Nẵng. (1978). Xã Luận: Xây dựng quan hệ sản xuất mới trong nông nghiệp [The editorial: Building new production relations in agriculture], *Quảng Nam-Đà Nẵng*, 14 June, p. 1.

Quảng Nam-Đà Nẵng. (1978). Vai trò của đảng viên trong đội sản xuất [The role of party members in production brigades], *Quảng Nam-Đà Nẵng*, 28 June, p. 2.

Quảng Nam-Đà Nẵng. (1978). Một số quy định về xây dựng hợp tác xã [Some regulations on establishing collectives], *Quảng Nam-Đà Nẵng*, 26 August, p. 1.

Quảng Nam-Đà Nẵng. (1978). Thăng Bình chuẩn bị xây dựng 10 hợp tác xã [Thăng Bình is about to establish 10 collectives], *Quảng Nam-Đà Nẵng*, 9 September, p. 2.

Quảng Nam-Đà Nẵng. (1978). Thành lập xong 98 hợp tác xã sản xuất nông nghiệp [98 agricultural collectives have been established], *Quảng Nam-Đà Nẵng*, 11 October, p. 3.

Quảng Nam-Đà Nẵng. (1978). Xã luận: Phấn đấu hoàn thành những mục tiêu về sản xuất và cải tạo nông nghiệp để trong năm 1976–1980 của tỉnh vào năm 1979 [The editorial: Do our best to meet 1976–1980 targets of production and agricultural transformation by 1979], *Quảng Nam-Đà Nẵng*, 21 October, p. 1.

Quảng Nam-Đà Nẵng. (1978). Đoàn cán bộ Ban cải tạo nông nghiệp Trung ương, các tỉnh Miền Trung và Hội liên hiệp phụ nữ Việt Nam thăm huyện Duy Xuyên [Cadres of Central Agricultural Transformation Committee, the Central Coast provinces and Vietnam Women's Union visited Duy Xuyên district], *Quảng Nam-Đà Nẵng*, 25 October, p. 1.

Quảng Nam-Đà Nẵng. (1978). Duy Xuyên khân trương xây dựng huyện để chỉ đạo và quản lý các hợp tác xã [Duy Xuyên district's rush to build capacity to lead collectives], *Quảng Nam-Đà Nẵng*, 25 October, p. 2.

Quảng Nam-Đà Nẵng. (1978). Nhìn vào đồng ruộng tập thể: Chủ nghĩa công điểm [Looking at collective fields: 'Work-pointism'], *Quảng Nam-Đà Nẵng*, 13 December, p. 2.

Quảng Nam-Đà Nẵng. (1978). Tổ chức lại sản xuất, phân công lại lao động nhằm phát triển và mở rộng lại ngành nghề sản xuất và kinh tế gia đình trong hợp tác xã nông nghiệp trên địa bàn huyện [Reorganising production and labour to facilitate development of handicrafts and household economy in the district], *Quảng Nam-Đà Nẵng*, 13 December, p. 1.

Quảng Nam-Đà Nẵng. (1979). 29-3-1975 – 29-3-1979: 4 năm lớn mạnh về kinh tế [From 29 March 1975 to 29 March 1979: 4 years of economic expansion], *Quảng Nam-Đà Nẵng*, 28 March, p. 1.

Quảng Nam-Đà Nẵng. (1979). Thành lập 32 hợp tác xã trong vụ Hè-Thu toàn tỉnh có 164 hợp tác xã [With 32 more collectives established, the province has 164 collectives for the summer–autumn crops], *Quảng Nam-Đà Nẵng*, 19 April, p. 1.

Quảng Nam-Đà Nẵng. (1979). Để đưa phong trào hợp tác xã nông nghiệp tiến lên mạnh mẽ và vững chắc [To speed up collectivisation forcefully and firmly], *Quảng Nam-Đà Nẵng*, 12 May, p. 1.

Quảng Nam-Đà Nẵng. (1979). Hợp tác xã Duy Phước ngọn cờ đầu của phong trào hợp tác hóa nông nghiệp tỉnh ta [Duy Phước is the leading collective in the province's agricultural cooperative movement], *Quảng Nam-Đà Nẵng*, 16 May, p. 1.

Quảng Nam-Đà Nẵng. (1979). Thăng Bình sơ kết hợp tác hóa nông nghiệp, phát động thi đua với HTX Duy Phước, Định Công và Vũ Thắng [A preliminary summing up of collectivisation in Thăng Bình], *Quảng Nam-Đà Nẵng*, 2 June, p. 2.

Quảng Nam-Đà Nẵng. (1979). Tăng cường công tác xây dựng đảng trong các hợp tác xã nông nghiệp [Intensifying building party organisation in the collectives], *Quảng Nam-Đà Nẵng*, 6 June, p. 1.

Quảng Nam-Đà Nẵng. (1979). HTX Bình Lãnh từ yếu kém vươn lên tiên tiến [Bình Lãnh collective is moving away from a position of weakness], *Quảng Nam-Đà Nẵng*, 9 June, p. 2.

Quảng Nam-Đà Nẵng. (1979). Cùng với cả Miền Nam tỉnh ta khẩn trương hoàn thành hợp tác hóa nông nghiệp [Our province, together with southern provinces, hurries to complete collectivisation], *Quảng Nam-Đà Nẵng*, 27 June, p. 1.

Quảng Nam-Đà Nẵng. (1979). Kết quả và kinh nghiệm phát huy quyền làm chủ tập thể ở HTX sản xuất nông nghiệp 1 Điện Nam [Result of and experiences from facilitating collective mastery in Điện Nam Collective No. 1], *Quảng Nam-Đà Nẵng*, 27 June, p. 3.

Quảng Nam-Đà Nẵng. (1979). Tiến hành đợt học tập trong các hợp tác xã nông nghiệp [Undertaking criticism and self-criticism in collectives], *Quảng Nam-Đà Nẵng*, 28 July, p. 2.

Quảng Nam-Đà Nẵng. (1979). Các biện pháp cấp bách đẩy mạnh sản xuất lương thực, thực phẩm ổn định đời sống nhân dân [Some urgent measures to increase food production], *Quảng Nam-Đà Nẵng*, 15 September, p. 1.

Quảng Nam-Đà Nẵng. (1979). Huyện Điện Bàn có kết quả học tập phát huy quyền làm chủ của xã viên trong hợp tác xã nông nghiệp [A preliminary summing up of undertaking criticism and self-criticism in collectives], *Quảng Nam-Đà Nẵng*, 15 September, p. 2.

Quảng Nam-Đà Nẵng. (1979). Nhận thức đúng đắn và thi hành nghiêm chỉnh việc tạm giao đất chuyên trồng màu cho xã viên sản xuất [Understanding well and seriously implementing a temporary redistribution of secondary land to members], *Quảng Nam-Đà Nẵng*, 19 September, p. 1.

Quảng Nam-Đà Nẵng. (1979). Xã viên làm chủ phát triển một đối tượng phá hoại hợp tác xã nông nghiệp [Members discover a pilferer in a collective], *Quảng Nam-Đà Nẵng*, 6 October, p. 2.

Quảng Nam-Đà Nẵng. (1979). Năm 1979 tỉnh ta căn bản hoàn thành hợp tác hóa nông nghiệp ở các huyện đồng bằng [The midlands of our province have completed collectivisation in 1979], *Quảng Nam-Đà Nẵng*, 17 October, p. 1.

Quảng Nam-Đà Nẵng. (1979). Huyện Thăng Bình tổng kết 2 năm cải tạo nông nghiệp [Thăng Bình district summing up 2 years of agricultural transformation], *Quảng Nam-Đà Nẵng*, 7 November, p. 2.

Quảng Nam-Đà Nẵng. (1979). Cuối năm 1978: Ra đời 107 hợp tác xã bao gồm 96,704 nông dân, chiếm 50% số hộ trong tỉnh [By late 1978: 107 cooperatives were established, including 96,704 farmers, accounting for 50 per cent of households in the province], *Quảng Nam-Đà Nẵng*, 2 December, p. 1.

Quảng Nam-Đà Nẵng. (1979). Hội nghị Ban cải tạo nông nghiệp tỉnh: Ra sức củng cố HTX để làm tốt vụ sản xuất Đông–Xuân [Provincial Committee for Agricultural Transformation: Strengthening cooperatives to make good in winter–spring production], *Quảng Nam-Đà Nẵng*, 26 December, p. 1.

Quảng Nam-Đà Nẵng. (1981). Quyết tâm đưa cuộc đấu tranh chống tiêu cực trong năm 1981 lên thành cao trào quần chúng, đều khắp vững chắc [Be resolute in fighting 'negativism' comprehensively in 1981], *Quảng Nam-Đà Nẵng*, 14 January, p. 1.

Quảng Nam-Đà Nẵng. (1981). Hợp tác xã Quế Tân I: Xây dựng con người, xây dựng hợp tác xã [Quế Tân Collective No. 1: Training people and building the collective], *Quảng Nam-Đà Nẵng*, 21 January, p. 2.

Quảng Nam-Đà Nẵng. (1981). Nhìn vào đồng ruộng tập thể: Giống lúa [Looking at collective fields: Rice seeds], *Quảng Nam-Đà Nẵng*, 28 February, p. 2.

Quảng Nam-Đà Nẵng. (1981). Bàn về công tác khoán sản phẩm cuối cùng đến người lao động trong sản xuất nông nghiệp [Discussing the assignment of final products to labourers in agricultural production], *Quảng Nam-Đà Nẵng*, 18 March, p. 1.

Quảng Nam-Đà Nẵng. (1981). Các hợp tác xã khẩn trương thực hiện thử khoán sản phẩm cuối cùng cho người lao động [Collectives must hurry in implementing the product contract], *Quảng Nam-Đà Nẵng*, 22 April, p. 1.

Quảng Nam-Đà Nẵng. (1981). Qua các hợp tác xã nông nghiệp làm thử việc khoán sản phẩm cuối cùng đến người lao động [Results of experimenting with the product contract], *Quảng Nam-Đà Nẵng*, 12 May, p. 2.

Quảng Nam-Đà Nẵng. (1981). Nhìn vào đồng ruộng tập thể: Lại chuyện chung và riêng [Looking at collective fields: Collective interest versus individual interest], *Quảng Nam-Đà Nẵng*, 23 May, p. 2.

Quảng Nam-Đà Nẵng. (1981). Qua các hợp tác xã làm thử việc khoán sản phẩm cuối cùng đến người lao động [An evaluation of the performance of collectives adopting the product contract], *Quảng Nam-Đà Nẵng*, 23 May, p. 1.

Quảng Nam-Đà Nẵng. (1981). Vụ Đông–Xuân 1980–1981 được mùa cả lúa và màu [Good rice and secondary crops harvested in winter–spring of 1980–1981], *Quảng Nam-Đà Nẵng*, 23 May, p. 1.

Quảng Nam-Đà Nẵng. (1981). Qua việc thực hiện ba khoán cho đội sản xuất trong vụ Đông–Xuân [An evaluation of implementing the three contracts for brigades in the winter–spring], *Quảng Nam-Đà Nẵng*, 30 May, p. 1.

Quảng Nam-Đà Nẵng. (1981). Phạm Đức Nam: Tích cực thực hiện khoán sản phẩm cuối cùng đến người lao động trong nông nghiệp [Phạm Đức Nam: Be positive in implementing the product contract], *Quảng Nam-Đà Nẵng*, 1 July, p. 1.

Quảng Nam-Đà Nẵng. (1981). Hợp tác xã Đại Minh khoán sản phẩm đến người lao động [Đại Minh collective has adopted the product contract], *Quảng Nam-Đà Nẵng*, 4 July, p. 2.

Quảng Nam-Đà Nẵng. (1981). Thư xã viên: Cách khoan mới ở quê tôi [Member's letter: New method of contracts in my village], *Quảng Nam-Đà Nẵng*, 4 July, p. 3.

Quảng Nam-Đà Nẵng. (1981). Hội nghị khoán sản phẩm trong hợp tác xã nông nghiệp ven biển Trung Trung Bộ và các tỉnh Tây Nguyên [A conference on the product contract in the Central Coast and Central Highlands collectives], *Quảng Nam-Đà Nẵng*, 8 July, p. 1.

Quảng Nam-Đà Nẵng. (1981). Khoán sản phẩm trên đất màu để làm vụ Đông–Xuân tốt nhất [Making contracts on secondary-crop land for the best winter–spring crop], *Quảng Nam-Đà Nẵng*, 15 August, p. 1.

Quảng Nam-Đà Nẵng. (1981). Ban Nông Nghiệp Tỉnh Ủy: Mấy kinh nghiệm khoán sản phẩm cuối cùng của Đại Minh [Provincial Committee of Agriculture: Some experiences from using the product contract in Đại Minh], *Quảng Nam-Đà Nẵng*, 19 August, p. 1.

Quảng Nam-Đà Nẵng. (1981). Huyện Thăng Bình phấn đấu đạt 65000 tấn lương thực năm 1981 [Thăng Bình is striving to produce 65,000 tonnes of food], *Quảng Nam-Đà Nẵng*, 9 September, p. 2.

Quảng Nam-Đà Nẵng. (1981). Một vài cách vận dụng khoán sản phẩm cuối cùng về cây lúa ở hợp tác xã Tam Ngọc [Application of the product contract for rice fields in Tam Ngọc collectives], *Quảng Nam-Đà Nẵng*, 12 September, p. 2.

Quảng Nam-Đà Nẵng. (1981). Phạm Đức Nam: Phát huy thắng lợi bước đầu mở rộng khoán sản phẩm cuối cùng ở tất cả hợp tác xã nông nghiệp cả tỉnh [Phạm Đức Nam: Extending the product contract to the rest of the collectives], *Quảng Nam-Đà Nẵng*, 23 September, p. 1.

Quảng Nam-Đà Nẵng. (1981). Các hợp tác xã ở Tam Kỳ, Thăng Bình, Tiên Phước căn bản hoàn thành khoán sản phẩm vụ Đông–Xuân [Collectives in Tam Kỳ, Thăng Bình and Tiên Phước have completed the adoption of the product contract in the winter–spring crop], *Quảng Nam-Đà Nẵng*, 28 October, p. 1.

Quảng Nam-Đà Nẵng. (1981). Mặt trận sản xuất nông nghiệp: Thành tựu của năm 1981 và nhiệm vụ vụ Đông–Xuân 1981–1982 [Agricultural production: The achievements of 1981 and the ongoing tasks for winter–spring 1981–1982], *Quảng Nam-Đà Nẵng*, 28 October, p. 1.

Quảng Nam-Đà Nẵng. (1981). Tổng kết sản xuất nông nghiệp năm 1981 và phát động thi đua giành vụ Đông–Xuân 1981–1982 thắng lợi toàn diện, vượt bậc [Summing up agricultural production in 1981 and calling for high achievements in the winter–spring of 1981–1982], *Quảng Nam-Đà Nẵng*, 4 November, p. 1.

Quảng Nam-Đà Nẵng. (1981). Mấy vấn đề cần chú ý trong việc khoán sản phẩm cuối cùng đến người lao động [Some notes about using the product contract], *Quảng Nam-Đà Nẵng*, 7 November, p. 1.

Quảng Nam-Đà Nẵng. (1981). Nân cao chất lượng khoán sản phẩm trong sản xuất Đông–Xuân [Improving the product contract in the winter–spring crop], *Quảng Nam-Đà Nẵng*, 25 November, p. 1.

Quảng Nam-Đà Nẵng. (1981). Chống hao hụt mất mát sản phẩm nông nghiệp khi thu hoạch [Preventing loss of collective produce during harvesting], *Quảng Nam-Đà Nẵng*, 1 December, p. 2.

Quảng Nam-Đà Nẵng. (1983). Hợp tác xã Duy Phước chặng đường 5 năm của phong trào hợp tác hóa nông nghiệp [Duy Phước collective over the past 5 years], *Quảng Nam-Đà Nẵng*, 5 March, p. 2.

Quảng Nam-Đà Nẵng. (1983). Củng cố và đưa các hợp tác xã nông nghiệp của huyện Tam Kỳ tiếp tục tiến lên [Solidifying and advancing collectives in Tam Kỳ district], *Quảng Nam-Đà Nẵng*, 4 June, p. 3.

Quảng Nam-Đà Nẵng. (1983). Hôm qua 16 tháng 4: Khai mạc hội nghị tổng kết phong trào hợp tác hóa nông nghiệp tỉnh ta [Yesterday, 16 April: The conference to sum up the province's agricultural collectivisation was opened], *Quảng Nam-Đà Nẵng*, 15 June, p. 1.

Quảng Nam-Đà Nẵng. (1983). Phong trào hợp tác hóa nông nghiệp: Sự kiện và con số [Overview of collectivisation: Events and figures], *Quảng Nam-Đà Nẵng*, 15 June, p. 2.

Quảng Nam-Đà Nẵng. (1983). Xã luận: Củng cố và phát triển quan hệ sản xuất mới trong nông nghiệp [The editorial: Solidifying and improving new production relations in agriculture], *Quảng Nam-Đà Nẵng*, 15 June, p. 1.

Quảng Nam-Đà Nẵng. (1983). Phạm Đức Nam: Công tác trước mắt để củng cố và phát triển quan hệ sản xuất mới ở nông thôn [Phạm Đức Nam: Ongoing tasks for solidifying and improving new production relations in rural areas], *Quảng Nam-Đà Nẵng*, 18 June, p. 1.

Quảng Nam-Đà Nẵng. (1983). Tam Phước củng cố hợp tác xã nông nghiệp [Tam Phước solidifies collectives], *Quảng Nam-Đà Nẵng*, 9 July, p. 3.

Quảng Nam-Đà Nẵng. (1983). Quế Sơn tổng kết phong trào hợp tác xã và củng cố hợp tác xã nông nghiệp [Quế Sơn sums up collectivisation and solidification of collectives], *Quảng Nam-Đà Nẵng*, 24 August, p. 3.

Quảng Nam-Đà Nẵng. (1983). Diễn biến sản lượng lúa cả tỉnh qua các năm [Paddy production over the past years], *Quảng Nam-Đà Nẵng*, 14 September, p. 1.

Quảng Nam-Đà Nẵng. (1983). Phạm Đức Nam: Kết quả năm 1983 và phương hướng phấn đấu năm 1984 trên mặt trận sản xuất nông nghiệp của tỉnh nhà [Phạm Đức Nam: The results of agricultural production in 1983 and plans for 1984], *Quảng Nam-Đà Nẵng*, 15 October, p. 1.

Quảng Nam-Đà Nẵng. (1983). Vụ Đông–Xuân 1983–1984 Thăng Bình củng cố hợp tác xã gắn liền với tập trung chỉ đạo vùng lúa có sản lượng cao [Thăng Bình will solidify the collective and extend high-yielding rice in the winter–spring of 1983–1984], *Quảng Nam-Đà Nẵng*, 9 November, p. 2.

Quảng Nam-Đà Nẵng. (1983). Xã luận: Củng cố hợp tác xã vấn để cấp bách đưa sản xuất nông nghiệp lên một bước [The editorial: Solidifying collectives is an urgent task to advance agriculture], *Quảng Nam-Đà Nẵng*, 14 November, p. 1.

Quảng Nam-Đà Nẵng. (1983). Củng cố và xác lập chế độ sở hữu tập thể trong hợp tác xã [Solidifying collective ownership], *Quảng Nam-Đà Nẵng*, 7 December, p. 1.

Quảng Nam-Đà Nẵng. (1984). Cuộc đấu tranh giữa hai con đường đang diễn ra ở một hợp tác xã [The struggle between two paths: Cooperative and individual farming in a collective], *Quảng Nam-Đà Nẵng*, 4 January, p. 4.

Quảng Nam-Đà Nẵng. (1984). Chung quanh vấn đề sử dụng đất nông nghiệp [The problem of using agricultural land], *Quảng Nam-Đà Nẵng*, 15 August, p. 2.

Quảng Nam-Đà Nẵng. (1984). Hội nghị tổng kết sản xuất nông nghiệp, phát động chiến dịch sản xuất vụ Đông–Xuân [A conference summing up five years of agricultural production and campaigning for the winter–spring crop], *Quảng Nam-Đà Nẵng*, 30 October, p. 1.

Quảng Nam-Đà Nẵng. (1984). Khoán sản phẩm cuối cùng đến người lao động những vướng mắc và cách giải quyết [The product contract: Problems and solutions], *Quảng Nam-Đà Nẵng*, 8 November, p. 1.

Quảng Nam-Đà Nẵng. (1985). Quán triệt nghị quyết hội nghị lần thứ 7 Ban chấp hành trung ương đảng: Ban chấp hành đảng bộ tỉnh quy định phương hướng nhiệm vụ năm 1985 [Full resolution of the 7th Plenum of the Provincial Party Committee: Plans for the year 1985], *Quảng Nam-Đà Nẵng*, 2 February, p. 1.

Quảng Nam-Đà Nẵng. (1985). Thăng Bình kỷ niệm 10 năm chiến thắng: Phong trào toàn dân làm thủy lợi, đẩy mạnh sản xuất nông nghiệp và kinh tế trong cả huyện [Thăng Bình celebrates 10th liberation anniversary: The people promote irrigation, agricultural production and the economy in the district], *Quảng Nam-Đà Nẵng*, 26 March, p. 3.

Quảng Nam-Đà Nẵng. (1985). Cần quản lý và sử dụng đất nông nghiệp một cách hợp lý [The need to use agricultural land rationally], *Quảng Nam-Đà Nẵng*, 25 April, p. 2.

Quảng Nam-Đà Nẵng. (1985). Hội nghị quản lý ruộng đất của tỉnh sử dụng tài nguyên đất với hiệu quả kinh tế cao nhất, chấm dứt việc cấp đất trái phép, xử lý nghiêm khắc những vụ lấn chiếm đất trái phép của nhà nước và tập thể [Provincial land management conference promotes land resources with the highest economic efficiency, terminates illegal land allocation, strictly handles illegal encroachment on state and collective land], *Quảng Nam-Đà Nẵng*, 25 April, p. 1.

Quảng Nam-Đà Nẵng. (1985). Tổng kết 3 năm thực hiện khoán sản phẩm đến người lao động trong nông nghiệp (1981–1984) [Summary of three years of implementing the product contract (1981–1984)], *Quảng Nam-Đà Nẵng*, 6 July, p. 1.

Quảng Nam-Đà Nẵng. (1985). Tổng kết sản xuất nông nghiệp năm 1985, chuẩn bị cho vụ Đông–Xuân tới [Summary of agricultural production in 1985 and preparing for the winter–spring crop], *Quảng Nam-Đà Nẵng*, 21 September, p. 1.

Quảng Nam-Đà Nẵng. (1985). Tại sao tiến độ huy động lương thực ở Thăng Bình chậm? [Why is food procurement in Thăng Bình slow?], *Quảng Nam-Đà Nẵng*, 7 December, p. 3.

Quảng Nam-Đà Nẵng. (1986). Năm năm phát triển sản xuất nông nghiệp [Five years of agricultural production], *Quảng Nam-Đà Nẵng*, 1 February, p. 2.

Quảng Nam-Đà Nẵng. (1986). Trách nhiệm của ngành lương thực trong việc để hao hụt một số khối lượng rất lớn lương thực [State food agencies need to take responsibility for considerable loss of staple food], *Quảng Nam-Đà Nẵng*, 16 August, p. 2.

Quảng Nam-Đà Nẵng. (1986). Củng cố và hoàn thiện công tác khoán sản phẩm cuối cùng đến người lao động trong nông nghiệp [Improving and perfecting the product contract], *Quảng Nam-Đà Nẵng*, 25 October, p. 1.

Quảng Nam-Đà Nẵng. (1986). Thực sự coi nông nghiệp là mặt trận hàng đầu [Agricultural sector needs to be regarded as top national priority], *Quảng Nam-Đà Nẵng*, 27 November, p. 1.

Quảng Nam-Đà Nẵng. (1987). Vì sao năng suất Đông-Xuân ở Thăng Bình giảm sút? [Why did rice productivity in Thăng Bình go down?], *Quảng Nam-Đà Nẵng*, 28 April, p. 2.

Quảng Nam-Đà Nẵng. (1987). Hội nghị củng cố và tăng cường quan hệ sản xuất trong nông nghiệp kết thúc tốt đẹp [The conference on solidifying agricultural production relations produced good results], *Quảng Nam-Đà Nẵng*, 16 June, p. 1.

Quảng Nam-Đà Nẵng. (1987). Suy nghĩ về Bình Lãnh: Aự giàu có còn ở phía trước [Think of Bình Lãnh: Prosperity is still ahead], *Quảng Nam-Đà Nẵng*, 23 July, p. 3.

Quảng Nam-Đà Nẵng. (1987). Nhìn vào đồng ruộng tập thể: Hai bàn cân ở hợp tác xã Điện Phước 2 [Looking at collective fields: Two different weighing scales at Điện Phước Collective No. 2], *Quảng Nam-Đà Nẵng*, 8 August, p. 2.

Quảng Nam-Đà Nẵng. (1987). Thăng Bình mở rộng hội nghị củng cố phong trào hợp tác hóa [Thăng Bình held a conference on solidifying collectives], *Quảng Nam-Đà Nẵng*, 18 August, p. 3.

Quảng Nam-Đà Nẵng. (1987). Xã luận: Thực hiện rộng rãi và đầy đủ nền dân chủ xã hội chủ nghĩa [The editorial: Fully implementing the socialist democracy], *Quảng Nam-Đà Nẵng*, 1 September, p. 1.

Quảng Nam-Đà Nẵng. (1987). Sơ kết sản xuất nông nghiệp năm 1987, chuẩn bị vụ sản xuất Đông–Xuân tới [Preliminary summing up of 1987 agricultural production and preparing for the winter–spring crop], *Quảng Nam-Đà Nẵng*, 17 September, p. 1.

Quảng Nam-Đà Nẵng. (1987). Điều tra nợ lương thực: Vấn đề giải quyết lương thực hiện nay [Investigation of food debt: How to deal with food problems], *Quảng Nam-Đà Nẵng*, 29 October, p. 2.

Quảng Nam-Đà Nẵng. (1987). Tổ chức thanh tra các cấp tăng cường công tác thanh tra, kiểm tra nhanh chóng phát hiện những vụ tiêu cực [Intensifying investigations of negativism], *Quảng Nam-Đà Nẵng*, 14 November, p. 1.

Quảng Nam-Đà Nẵng. (1987). Hợp tác xã Duy Thành từng bước hoàn thiện khoán sản phẩm đối với cây lúa [Duy Thành Collective gradually perfected the product contract], *Quảng Nam-Đà Nẵng*, 17 December, p. 3.

Quảng Nam-Đà Nẵng. (1987). Hợp tác xã Điện Nam 2 khoán mới động lực mới [New farming arrangements created new incentives in Điện Nam Collective No. 2], *Quảng Nam-Đà Nẵng*, 24 December, p. 2.

Quảng Nam-Đà Nẵng. (1987). Chuyện đồng ruộng cuối năm [Collective farming at the end of the year], *Quảng Nam-Đà Nẵng*, 31 December, p. 3.

Quảng Nam-Đà Nẵng. (1988). Hoàn thiện cơ chế khoán sản phẩm trong nông nghiệp [Perfecting the product contract system], *Quảng Nam-Đà Nẵng*, 1 March, p. 1.

Quảng Nam-Đà Nẵng. (1988). Bình Triều 2 qua vụ Đông–Xuân 1987–1988 [Performance of Bình Triều Collective No. 2 in the winter-spring of 1987–1988], *Quảng Nam-Đà Nẵng*, 10 May, p. 4.

Quảng Nam-Đà Nẵng. (1988). Sự thật về cách khoán mới ở Bình Tú 1 [The true story of new contract arrangements in Bình Tú Collective No. 1], *Quảng Nam-Đà Nẵng*, 23 June, p. 4.

Quảng Nam-Đà Nẵng. (1988). Qua một năm cải tiến công tác khoán sản phẩm trong sản xuất nông nghiệp [An evaluation after one year of improving the product contract system], *Quảng Nam-Đà Nẵng*, 30 August, p. 2.

Quang Truong. (1987). Agricultural collectivization and rural development in Vietnam: A north/south study (1955–1985). PhD thesis, Vrije Universiteit te Amsterdam, Amsterdam.

Sài Gòn Giải Phóng. (1976). Phạm Hùng: Miền Nam có trách nhiệm lớn đối với sự nghiệp cách mạng xã hội chủ nghĩa chung cả nước [Phạm Hùng: The south has a big responsibility for the country's socialist revolution], *Sài Gòn Giải Phóng*, 1 July.

Scott, J. C. (1977). *The Moral Economy of the Peasant: Rebellion and Subsistence in Southeast Asia*. New Haven, CT: Yale University Press.

Scott, J. C. (1998). *Seeing Like a State: How Certain Schemes to Improve the Human Condition Have Failed*. New Haven, CT: Yale University Press.

Scott, J. C. and Kerkvliet, B. J. (1986). *Everyday Forms of Peasant Resistance in South-East Asia*. Hove, UK: Psychology Press.

Selden, M. (1994). Pathways from collectivization: Socialist and post-socialist agrarian alternatives in Russia and China. *Review (Fernand Braudel Center)* 17(4): 423–49.

Shue, V. (1990). *The Reach of the State: Sketches of the Chinese Body Politic*. Stanford, CA: Stanford University Press.

Shue, V. (1994). State power and social organization in China. In J. Migdal, A. K. Samuel and V. Shue (eds), *State Power and Social Forces: Domination and Transformation in the Third World*. Cambridge: Cambridge University Press. doi.org/10.1017/CBO 9781139174268.006.

Sikor, T. (2001). The allocation of forestry land in Vietnam: Did it cause the expansion of forests in the northwest? *Forest Policy and Economics* 2(1): 1–11. doi.org/10.1016/S1389-9341(00)00041-1.

Sở Nông Nghiệp Phát Triển Nông Thôn Quảng Nam. (2005). *Kết quả sản xuất nông nghiệp năm 1976–2004* [*Agricultural Production 1976–2004*]. Tam Kỳ: Sở Nông Nghiệp và Phát Triển Nông Thôn Quảng Nam.

Sở Thông Tin Văn Hóa An Giang (STTVHAG). (1978). *Thông tin phổ thông* [*General Information*]. Vol. 9. Long Xuyên: NXB Sở Thông Tin Văn Hóa An Giang.

Sobhan, R. (1993). *Agrarian Reform and Social Transformation: Preconditions for Development*. London: Zed Books.

Sokolovsky, J. (1990). *Peasants and Power: State Autonomy and the Collectivization of Agriculture in Eastern Europe*. Westview Special Studies on the Soviet Union and Eastern Europe. Boulder, CO: Westview Press.

Taylor, P. (2001). *Fragments of the Present: Searching for Modernity in Vietnam's South*. Honolulu: University of Hawai'i Press.

Taylor, P. (2004). Introduction: Social inequality in a socialist state. In P. Taylor (ed.), *Social Inequality in Vietnam and the Challenges to Reform*. Singapore: Institute of Southeast Asian Studies.

Thayer, C. A. (1978). Dilemmas of development in Vietnam. *Current History* (December): 221–5.

Thế Đạt. (1981). *Nền nông nghiệp Việt Nam từ sau cách mạng tháng Tám năm 1945* [*Vietnamese Agriculture Since the August Revolution 1945*]. Hà Nội: NXB Nông Nghiệp.

Thompson, L. C. (2010). *Refugee Workers in the Indochina Exodus, 1975–1982*. Jefferson, NC: McFarland & Co.

Tỉnh Ủy Quảng Nam (TUQN). (2003). *Quảng Nam Anh Hùng, thời đại Hồ Chí Minh, Kỷ Yếu 6/2003* [*Quảng Nam is a Hero in the Age of Hồ Chí Minh*]. Tam Kỳ: Tỉnh Ủy Quảng Nam.

Tỉnh Ủy Quảng Nam-Đà Nẵng. (1983). Nghị Quyết 09/NQ-TV (ngày 5-11-1983) về việc củng cố và hoàn thiện quan hệ sản xuất xã hội chủ nghĩa [Provincial Party Committee's Resolution No. 09/NQ-TV (5 November 1983) on solidifying and perfecting production relations in agriculture], *Tỉnh Ủy Quảng Nam-Đà Nẵng*, 11 November, p. 1.

Tỉnh Ủy Quảng Nam-Đà Nẵng. (1987). Nghị quyết số 03/NQ-TU: Tiếp tục củng cố và tăng cường quan hệ sản xuất, hòa thiện cơ chế sản phẩm trong nông nghiệp [Resolution No. 03/NQ-TU: Continue to improve and perfect the product contract], *Tỉnh Ủy Quảng Nam-Đà Nẵng*, 22 June.

Tỉnh Ủy Quảng Nam-Đà Nẵng. (1987). Nghị quyết của Tỉnh ủy tiếp tục củng cố và tăng cường quan hệ sản xuất, hoàn thiện cơ chế khoán sản phẩm [Provincial Party Committee's resolution on continuing to solidify production relations and perfect the product contract], *Tỉnh Ủy Quảng Nam-Đà Nẵng*, 9 July, p. 1.

Tố Hữu. (1979). Phát động phong trào quần chúng thực hiện thắng lợi công cuộc cải tạo xã hội chủ nghĩa đối với nông nghiệp Miền Nam [Campaign to succeed in socialist agricultural transformation in the south]. In Võ Chí Công and Tố Hữu (eds), *Khẩn trương và tích cực đẩy mạnh phong trào hợp tác hóa nông nghiệp Miền Nam* [*Urgently and Positively Promote the Acceleration of Collectivisation in the South*]. Hà Nội: NXB Sự Thật.

Tô Thành Tâm. (1990). Vấn đề ruộng đất và hợp tác hóa nông nghiệp ở An Giang [Land and collectivisation issues in An Giang], *Thông Tin Lý Luận*, 8 August, p. 8.

Trần Hữu Đính. (1994). *Quá trình biến đổi về chế độ sở hữu và cơ cấu giai cấp nông thôn Đồng Bằng Sông Cửu Long (1969–1975)* [*The Process of Ownership and Class Structure Change in Rural Mekong Delta, 1969–1975*]. Hà Nội: NXB Khoa Học Xã Hội.

Trần Ngọc Cư-Ban Nông Nghiệp Tỉnh QN-DN. (1984). Kinh tế gia đình ở tỉnh ta [The household economy in our province], *Quảng Nam-Đà Nẵng*, 29 November.

Trần Phương. (1968). *Cách mạng ruộng đất ở Việt Nam [Land Revolution in Vietnam]*. Hà Nội: NXB Khoa Học Xã Hội.

Trần Văn Doãn. (1986). *Như thế nào là nông nghiệp một bước lên sản xuất lớn xã hội chủ nghĩa [What is One Step of Agriculture Towards Socialist Large-Scale Production]*. Hà Nội: NXB Nông Nghiệp.

Tuma, E. H. (1965). *Twenty-Six Centuries of Agrarian Reform: A Comparative Analysis*. Berkeley, CA: University of California Press.

Ủy Ban Nhân Dân Tỉnh An Giang (UBNDTAG). (1989). *Quyết định 303/QĐ-UB [An Giang People's Committee Directive No. 303/QĐ-UB]*, 4 October. Long Xuyên: An Giang.

Ủy Ban Nhân Dân Tỉnh An Giang (UBNDTAG). (2003). *Địa Chí An Giang [An Giang Province]*. An Giang: Ủy Ban Nhân Dân Tỉnh An Giang.

Vasavakul, T. (1993). Vietnam: Sectors, classes, and the transformation of a Leninist state. In J. W. Morley (ed.), *Driven by Growth: Political Change in the Asia-Pacific Region*. Armonk, NY: M. E. Sharpe.

Võ Chí Công. (1978). Con đường làm ăn tập thể của nông dân [The collective farmer's way]. In Võ Chí Công, Nguyễn Thành Thơ, Phan Văn Đáng and Phạm Văn Kiết (eds), *Con đường làm ăn tập thể của nông dân [The Collective Farmer's Way]*. Hồ Chí Minh: NXB Tp. Hồ Chí Minh.

Vo Nhan Tri. (1990). *Vietnam's Economic Policy Since 1975*. Singapore: Institute of Southeast Asian Studies.

Võ Tòng Xuân and Chu Hữu Quý. (1994). *Đề Tài KX 08-11: Tổng kết khoa học phát triển tổng hợp kinh tế xã hội nông thôn qua 7 năm xây dựng và phát triển An Giang [KX Account 08-11: Summing Up An Giang's Socioeconomic Construction and Development over the Past 7 Years]*. Long Xuyên: Chương Trình Phát Triển Nông Thôn An Giang.

Võ Văn Kiệt. (1977). Cải tạo Xã hội chủ nghĩa nhiệm vụ hàng đầu của Thành phố Hồ Chí Minh [Socialist transformation is the top priority of Hồ Chí Minh City], *Đại Đoàn Kết*, 9 July.

Võ Văn Kiệt. (1985). *Thực hiện đồng bộ ba cuộc cách mạng ở nông thôn [Simultaneous Execution of Three Revolutions in Rural Areas]*. Hồ Chí Minh: NXB Tổng Hợp TP Hồ Chí Minh.

Vũ Huy Thúc. (1979). *Tìm hiểu chế độ ruộng đất Việt Nam nửa đầu thế kỷ XIX [Examining Vietnam's Land Tenure System in the First Half of the Nineteenth Century]*. Hà Nội: NXB Khoa Học Xã Hội.

Vũ Oanh. (1984). *Hoàn thành điều chỉnh ruộng đất, đẩy mạnh cải tạo xã hội chủ nghĩa đối với nông nghiệp các tỉnh Nam Bộ [Completing Land Redistribution and Speeding Up Agricultural Transformation in the Southern Region]*. Hà Nội: NXB Sự Thật.

White, C. (1981). Agrarian reform and national liberation in the Vietnamese revolution 1920–1957. PhD thesis, Cornell University, Ithaca, NY.

White, C. P. (1988). Alternative approaches to the socialist transformation of agriculture in postwar Vietnam. In D. Marr and C. White (eds), *Postwar Vietnam: Dilemmas in Socialist Development*. Ithaca, NY: Cornell University Press.

Wiegersma, N. (1988). *Vietnam: Peasant Land, Peasant Revolution— Patriarchy and Collectivity in the Rural Economy*. New York: St Martin's Press. doi.org/10.1007/978-1-349-09970-2.

Womack, B. (1987). The party and the people: Revolutionary and postrevolutionary politics in China and Vietnam. *World Politics* 39(4): 479–507. doi.org/10.2307/2010289.

Xuân Thu and Quang Thiện. (2005). Đêm trước đổi mới: Công phá 'lũy tre' [On the eve of the renovation: Breaking through the 'bamboo hedges']. *Tuổi trẻ Online*, 14 December. Available from: tuoitre.vn/ tin/chinh-tri-xa-hoi/phong-su-ky-su/20051204/dem-truoc-doi-moi-cong-pha-luy-tre/111625.html (accessed 4 October 2017).

Zhou, K. X. (1996). *How the Farmers Changed China: Power of the People*. Boulder, CO: Westview Press.

www.ingramcontent.com/pod-product-compliance
Lightning Source LLC
Chambersburg PA
CBHW040820280326
41926CB00093B/4594